Springer Theses

Recognizing Outstanding Ph.D. Research

For further volumes:
http://www.springer.com/series/8790

Aims and Scope

The series "Springer Theses" brings together a selection of the very best Ph.D. theses from around the world and across the physical sciences. Nominated and endorsed by two recognized specialists, each published volume has been selected for its scientific excellence and the high impact of its contents for the pertinent field of research. For greater accessibility to non-specialists, the published versions include an extended introduction, as well as a foreword by the student's supervisor explaining the special relevance of the work for the field. As a whole, the series will provide a valuable resource both for newcomers to the research fields described, and for other scientists seeking detailed background information on special questions. Finally, it provides an accredited documentation of the valuable contributions made by today's younger generation of scientists.

Theses are accepted into the series by invited nomination only and must fulfill all of the following criteria

- They must be written in good English.
- The topic should fall within the confines of Chemistry, Physics, Earth Sciences, Engineering and related interdisciplinary fields such as Materials, Nanoscience, Chemical Engineering, Complex Systems and Biophysics.
- The work reported in the thesis must represent a significant scientific advance.
- If the thesis includes previously published material, permission to reproduce this must be gained from the respective copyright holder.
- They must have been examined and passed during the 12 months prior to nomination.
- Each thesis should include a foreword by the supervisor outlining the significance of its content.
- The theses should have a clearly defined structure including an introduction accessible to scientists not expert in that particular field.

Nouamane Laanait

Ion Correlations at Electrified Soft Matter Interfaces

Doctoral Thesis accepted by
the University of Illinois, Chicago, USA

Author
Dr. Nouamane Laanait
Chemical Sciences and Engineering
Argonne National Laboratory
Argonne, IL
USA

Supervisor
Prof. Mark L. Schlossman
Department of Physics
University of Illinois, Chicago
Chicago, IL
USA

ISSN 2190-5053
ISBN 978-3-319-00899-8
DOI 10.1007/978-3-319-00900-1
Springer Cham Heidelberg New York Dordrecht London

ISSN 2190-5061 (electronic)
ISBN 978-3-319-00900-1 (eBook)

Library of Congress Control Number: 2013940953

© Springer International Publishing Switzerland 2013
This work is subject to copyright. All rights are reserved by the Publisher, whether the whole or part of the material is concerned, specifically the rights of translation, reprinting, reuse of illustrations, recitation, broadcasting, reproduction on microfilms or in any other physical way, and transmission or information storage and retrieval, electronic adaptation, computer software, or by similar or dissimilar methodology now known or hereafter developed. Exempted from this legal reservation are brief excerpts in connection with reviews or scholarly analysis or material supplied specifically for the purpose of being entered and executed on a computer system, for exclusive use by the purchaser of the work. Duplication of this publication or parts thereof is permitted only under the provisions of the Copyright Law of the Publisher's location, in its current version, and permission for use must always be obtained from Springer. Permissions for use may be obtained through RightsLink at the Copyright Clearance Center. Violations are liable to prosecution under the respective Copyright Law.
The use of general descriptive names, registered names, trademarks, service marks, etc. in this publication does not imply, even in the absence of a specific statement, that such names are exempt from the relevant protective laws and regulations and therefore free for general use.
While the advice and information in this book are believed to be true and accurate at the date of publication, neither the authors nor the editors nor the publisher can accept any legal responsibility for any errors or omissions that may be made. The publisher makes no warranty, express or implied, with respect to the material contained herein.

Printed on acid-free paper

Springer is part of Springer Science+Business Media (www.springer.com)

To my father, Lahoussine, who opened my eyes to the wonders of Physics
To my mother, Menana, my inexhaustible source of courage
To my wife, Elizabeth, for her constant support and tender affection
To my son, Tristan, who inspires me to be better, to do more, to feel life in every breath

Supervisor's Foreword

This Ph.D. thesis by Nouamane Laanait explores the role of correlations between ions in determining the distribution of ions near a liquid/liquid interface. Ions are distributed near charged molecules and particles in solution. These include counterions that are released upon dissolving the molecules and particles or, possibly, other electrolytes dissolved in solution. The presence of these counter- and co-ions alters, or screens, the electrostatic interaction between the charged entities. As a result, the distribution of ions about charged molecules and particles in solution plays an important role in many scientific areas. Gouy and Chapman separately addressed this topic in the early twentieth century within the context of counter-ion distributions in solution near a charged electrode. Subsequently, Debye and Hückel studied the ion distribution about a single ion within an electrolyte solution. Both of these approaches were based upon a mean-field theory for which the distributed ions interact electrostatically with the mean field of all other ions. Except for modeling the solvent as a dielectric continuum, all other interactions were neglected. Since then, many interesting ideas have been introduced about the role of other interactions, including non-mean-field correlations and fluctuations. It has been suggested that non-mean-field effects are responsible for unusual effects, such as charge inversion of colloids and DNA condensation in solutions of multivalent ions. However, a direct connection had not been previously established between the predictions of these novel ion distributions and their measurement. Nouamane's Ph.D. research, described in this thesis, is the first to establish this connection. The National Science Foundation Division of Chemistry grants 0615929 and 0910825 supported his work.

Nouamane joined my research group in May of 2006, just a few years after Guangming Luo, a post-doctoral associate recruited from Beijing, had started to use X-ray surface scattering to probe the distribution of ions at a liquid/liquid interface. The X-ray experiments utilized the high-precision liquid surface scattering instrument situated at the ChemMatCARS sector of the Advanced Photon Source (Argonne National Laboratory), where Binhua Lin and Mati Meron ably assisted us in these experiments and in Nouamane's subsequent experiments. Guangming's research established the importance of considering ion-solvent interactions to understand ion distributions. Although several authors had previously introduced ion-solvent interactions into a Poisson–Boltzmann formalism to produce a theory of ion distributions, Daikhin, Kornyshev, and Urbakh's clear description of the role of a free energy profile on ion distributions at a liquid/liquid interface influenced us. Guangming's model of the free energy profile with a simple analytic form provided excellent fits to our data, but a more fundamental determination of the free energy profile awaited a conversation with Ilan Benjamin (University of California at Santa Cruz) in a study lounge at the Argonne National Laboratory Guest House. Subsequently, Ilan used molecular dynamics computer simulations to determine the potential of mean force for the relevant ions and Guangming showed that this description of ion-solvent interactions, within the context of a modified Poisson–Boltzmann equation, yielded ion distributions that matched his data.

Guangming's early measurements were limited by the small electric potential difference across the liquid/liquid interface that was achievable in a system for which the potential difference is established by ion partitioning across the interface. Nouamane's first experimental work in the summer of 2006, in collaboration with his fellow graduate students Binyang Hou and Jaesung Yoon, utilized a full electrochemical cell that Guangming and Petr Vanysek (Northern Illinois University) had designed prior to Guangming's departure from my lab. Along with the appropriate electrolytes, this cell provided a range of ± 400 mV in electric potential difference, which led to strikingly different X-ray reflectivity measurements than had been observed previously. Needless to say, this did not immediately lead to an understanding of the role of ion correlations in ion distributions, though it did provide an experimental method to measure ion distributions on the nanoscale under conditions in which the strength of ion correlations can be tuned. However, years of experimental and theoretical research were required to understand these data.

After establishing a protocol for measuring reproducible X-ray reflectivity from these samples, many ancillary experiments described or mentioned in this thesis were required for a full understanding of these measurements. In addition, several theoretical approaches to the analysis of these data were undertaken. The earliest analysis revealed that the monotonic potential of mean force, previously used successfully by Guangming, underestimated the interfacial ion concentration exhibited by the new data. Nouamane addressed this issue by adding a Gaussian well to the potential of mean force. This led to an understanding that organic anions ($TPFB^-$) were forming a condensed layer at the organic/water interface under large, positive electric potentials. It became apparent that this effect was more than just a perturbation on the Gouy–Chapman theory of ion distributions, and these results were published in the Journal of Chemical Physics in 2010.

In seeking a fundamental understanding of the condensed ionic layer, Nouamane pursued several different approaches. These included the very demanding task of a molecular dynamics simulation of the organic anion, $TPFB^-$, at the organic/water interface. This calculation involved the transfer of a significantly more complex ion than had been studied previously. Ilan Benjamin supervised Nouamane in this calculation, which involved parameter development, quantum computations, and improvements to the simulation methodology. The single-ion potential of mean force for TPFB$^-$ derived from these calculations demonstrated conclusively that ion-solvent interactions, by themselves, could not explain the X-ray data.

Nouamane then considered the effect of ion correlations. At first, he modified and extended the so-called Steric Poisson–Boltzmann (SPB) theory by Andelman and co-authors, which treats excluded volume correlations by considering the effect of solvent entropy. Although a promising description of Nouamane's X-ray data, this thesis constitutes the first publication of this work. At about the time that Nouamane was finishing the SPB analysis, he became very enthusiastic about the Debye–Hückel hole theory of electrostatic correlations. A recently formulated version of this theory, named after two mid-twentieth century scientists, is a modern approach to ion correlations that are the result of long-range electrostatic interactions. This is a very different approach to ion correlations than the excluded volume approach of SPB theory; however, it could not be applied by itself for a quantitative analysis of the X-ray data.

By combining the effects of ion-solvent interactions and ion correlations, Nouamane realized the breakthrough required to analyze his X-ray data on ion distributions. Nouamane showed that his data could be fit by a density functional approach that combined his MD simulation of ion-solvent interactions with the Debye–Hückel hole theory in a weighted density approximation. Fitting the X-ray data required only one free parameter, the interfacial roughness, whose fitted values agreed with those calculated from capillary wave theory. The agreement between X-ray data and ion distribution theory confirmed the long-standing prediction of a sharply localized double layer, which is a signature of ion correlation models. Application of this new approach to the analysis of X-ray data that probes ion distributions on the nanoscale is an important and original contribution to the scientific literature.

The density functional approach taken by Nouamane should have broad applicability. For example, as part of his thesis research, Nouamane applied this approach to describe an interfacial tension measurement of the variation of interfacial excess charge with electric potential difference. No adjustable parameters are required for the striking agreement of Nouamane's measurements of excess charge and his theoretical prediction. This last result demonstrated an important connection between X-ray measurements that probe ion distributions on the nanoscale and common electrochemical measurements that can be carried out in many laboratories. A short report of Nouamane's work on ion correlations was published in the Proceedings of the National Academy of Sciences in 2012.

On a personal note, I would like to thank Nouamane for the many enjoyable and exciting scientific interactions that we shared during his years working in my laboratory.

Chicago, April 2013   Prof. Mark L. Schlossman

Acknowledgments

I would like to thank my family for their encouragement and support throughout my graduate studies.

Special thanks to my former colleagues at the University of Illinois at Chicago, Jaesung Yoon, Binyang Hou, Miroslav Mihaylov, and Hao Yu for their contributions to this project.

I would like to thank Prof. Ilan Benjamin (University of California at Santa Cruz) for his mentorship in Molecular Dynamics Simulations.

I have had the immense pleasure and opportunity to learn from and work alongside my advisor, Prof. Mark L. Schlossman. The valuable investigations and results, that I hope to have demonstrated in this work, abound with his original ideas and insights. I am enormously grateful for his mentorship, and all the scientific training he offered or sponsored in collaborations with other researchers. Above all, I would like to thank Prof. Schlossman for encouraging me to think and work independently.

Contents

1	Introduction	1
	References	3
2	The Poisson–Boltzmann Equation	5
	References	8
3	Electrochemical Methods	9
	3.1 The Electrified Interface Formed by Two Immiscible Electrolyte Solutions	10
	3.2 The Potential of Zero Charge	16
	3.3 Conductivity Measurements of the Dissociation of BTPPATPFB	18
	References	23
4	X-ray Reflectivity Studies of Ion Condensation at the Electrified Liquid/Liquid Interface	25
	4.1 Experimental Procedure	26
	4.2 Data Analysis	28
	4.3 Concluding Remarks	36
	References	36
5	Sterically Modified Poisson–Boltzmann Equation	39
	5.1 Lattice Gas Approach	40
	5.2 Density Functional Approach	50
	5.3 Generalized SPB Theory	51
	5.4 Numerical Implementation of Electrostatic Boundary Value Problems	54
	5.5 Experimental Tests of the SPB Theory	58
	5.6 Concluding Remarks	64
	References	65

6	**Molecular Dynamics Simulation of Solvent Correlations**	67
	6.1 MD Simulation of a Liquid/Liquid Interface	68
	6.2 Potential of Mean Force of Na^+, Li^+, and Cl^-	70
	6.3 Potential of Mean Force of $TPFB^-$	77
	6.4 Concluding Remarks	83
	References	83
7	**The Role of Electrostatic Ion Correlations in Ion Condensation**	85
	7.1 The Debye–Hückel Hole Theory	86
	7.2 A Density Functional Theory of Ion Correlations	89
	7.3 Experimental Tests of the PB/MD/DHH Theory	92
	7.4 Conclusion	98
	References	99
Appendices		101

Abbreviations

BTPPA	Bis(TriPhenyl Phosphoranylidene) Ammonium
DFT	Density-Functional Theory
DHH	Debye–Hückel Hole
GSPB	Generalized Steric Poisson–Boltzmann
LDA	Local Density Approximation
LJ	Lennard-Jones
MC	Monte Carlo
MD	Molecular Dynamics
PB	Poisson–Boltzmann
PB-PMF	Poisson–Boltzmann with Potential of Mean Force
PMF	Potential of Mean Force
PZC	Potential of Zero Charge
SC	Strong Coupling Limit
SPB	Steric Poisson–Boltzmann
TPFB	Tetrakis(PentaFluorophenyl) Borate
WDA	Weighted Density Approximation

Chapter 1
Introduction

The study of ion distributions near charged surfaces holds great promise for the understanding of various biophysical processes [1–3], such as ion and electron transfer across biomembranes and DNA condensation to name but a few, and finds natural applications in the industry, most notably in energy storage in electrochemical capacitors [4]. The study of ion distributions is a multidisciplinary field, spanning statistical mechanics, many-body physics, computer simulation, electrochemistry, and physical chemistry. However, the origins of this field belong to electrochemistry, when Gouy and Chapman [5, 6], almost a century ago, tried to describe how ions are distributed near an electrode in a system akin to an electrochemical cell, and discovered the celebrated Poisson-Boltzmann equation (PB). This theory was able to successfully describe a number of properties of electrolytes, but was found to poorly predict the behavior of systems that are highly concentrated or near highly charged electrodes. These shortcomings prompted a number of researchers to modify some of the simplifying assumptions entering PB. For instance, Stern [7] introduced a solvent layer (commonly called the Stern layer) near the charged surface from which ions are excluded, and represents the first attempt to address solvent structure within a theory of the electrical double layer. A milestone in the study of ion distributions is the theory of bulk electrolytes by Debye and Hückel [8]. In their attempt to construct a self-consistent theory, they linearized the PB equation, thereby introducing a methodology to deal with screened interactions that found wide applicability in physics, and defined properties of solutions that are still widely used in chemistry. Debye-Hückel (DH) like theories, also constitute a class of models where ion correlations can be calculated exactly. Ever since the discovery of the Poisson-Boltzmann equation, researchers have tried to modify it to include a more realistic description of ion interactions in solution. In addition to the electrostatic interaction, which is the only interaction considered in the PB and DH theories, specific ion-solvent interactions such as hydration are very strong and cannot be neglected. Also, interactions due to ions having a finite size [9], which is ignored in the PB treatment, places severe restraints on the packing and layering structures of ion distributions near surfaces. There has been steady theoretical progress in investigating all of the above interactions within the framework of the PB equation, as well as in more sophisticated approaches. However, the measurement of the electrical double layer at an atomic or molecular level, has only become possible in the past two decades, starting with the x-ray standing waves measurements of the diffuse layer at a solid/liquid interface by Bedzyk [10]. In the past decade, x-ray scattering studies have been performed to measure ion density profiles near solid/liquid interfaces by Fenter and coworkers using resonant x-ray reflectivity [11], near DNA molecules using small-angle x-ray scattering by Pollack et al. [12], and at liquid/liquid interfaces by Schlossman's group [13–16]. These studies have revealed that the energetics of electrical double layers are highly dependent on the chemical nature of the ions, the specific ion-solvent interactions, and solvent structure near the charged surface. A theoretical model that aims to describe all these specific interactions will be hopelessly complicated and must be tailored to the problem at hand. Hence, severely limiting its applicability in other settings, and obscuring the physical picture it should provide of the salient features of the double layer. Fortunately, many of these specific interactions are easily addressed in computer simulations, specifically Molecular Dynamics simulations. Therefore, the general approach that we take in this work to interpret the experimental investigations, can be succinctly summed up as : Derive what can be derived, then simulate the rest.

A key aspect of ion distributions that has proven difficult to describe, eluding theoretical treatment and experimental investigation alike, is electrostatic ion correlations (see [3] for an excellent review). The latter have been shown to give rise to a number of counterintuitive results. Notorious among these is the phenomenon of *charge reversal* or *overcharging*. First discovered using Monte Carlo simulations (MC) of the electrical double layer (1:1 Restricted Primitive Model) by Torrie et al. [17], it was found that at high surface charge densities and concentrations, the ions strongly condensed on the charged plate exhibiting quasi-layering. This ionic layer seemed to overcharge the plate, as a result of the appearance of a potential drop. Subsequent electrophoresis experiments of colloids confirmed this result, where the mobility of a colloid in a suspension with high ionic strength and multivalent ions is reversed with respect to its bare charge. A succinct description of overcharging can be given if we consider ion correlations: Gain in electrostatic free energy due to correlations in the positions of the screening counter-ions. However this gain needs to be large enough to counteract the loss in entropy, hence the need for strong electrostatic interactions in a system. Of equal significance to the study of the electrical double layer is the discovery of like-charge attraction between two colloids in a suspension based on the hypernetted chain approximation by Patey [18] and eventually confirmed by simulations and experiments. This attraction is usually studied in the case of planar geometry, where an analytic solution of the Poisson-Boltzmann equation exists, for instance the attraction between two like-charged plates in a solution of counter-ions [19]. This like-charge attraction was proved to be impossible in the framework of mean-field theories such as Poisson-Boltzmann and DLVO theory (Derjaguin and Landau, Verwey and Overbeek) [20, 21]. It is widely believed that like-charge attraction is the mechanism responsible for fundamental biophysical processes, such as the formation of actin bundles and the aggregations of DNA, a phenomenon known as DNA condensation [22], and is responsible for the compact form of genes. Charge reversal and like-charge attraction demonstrated the need for theories and models, beyond the mean-field approximation, to describe ion correlations. Addressing the latter has shed much light on the stability and phase separation of colloidal suspensions [3]. However, the role that strong ion correlations play in DNA condensation remains an issue of contention.

In this work, the issue of describing ion correlations in charged soft matter is addressed using experimental investigations, computer simulations, and density functional theory. Synchrotron x-ray scattering reveals dense ionic condensation at the liquid/liquid interface, when the latter is polarized with an electric field (see Chap. 3). This observation is used as a stringent test of models of the electrical double layer. Specifically, systematic control of the magnitude of this ionic layer, through the interfacial electric field, allows for a detailed study of correlations as a function of the Coulomb coupling strength in the system. A mean-field description of ion interactions, the Poisson-Boltzmann equation (see Chap. 2), is incapable of predicting physical density profiles when the ions are strongly correlated, this is demonstrated by comparing the predictions of the Poisson-Boltzmann equation to x-ray reflectivity data, which probes the electron density profile with a molecular resolution (see Chap. 4). A coarse-grained Poisson-Boltzmann theory based on a lattice Coulomb gas removes the divergences of the Poisson-Boltzmann equation, and produces properties of the electrical double layer in modest agreement with thermodynamic measurements of the interfacial charge (see Chap. 5). Nonetheless, significant deviations between theory and x-ray data remain at the highest probed potentials. A density functional theory is proposed that suggests that the observed ionic condensation is driven by electrostatic ion correlations. The latter are described by a nonlocal free energy functional based on the Debye-Hückel Hole theory of a one-component plasma. Once corrected for specific ion-solvent interactions at the interface that are mapped out using Molecular Dynamics simulations (see Chap. 6), remarkable agreement between the predicted density profiles and the structural measurements is found, without any adjustable parameters in the theoretical model. The proposed density functional also produces global electrostatic properties of the double layer in excellent agreement with the thermodynamic data (see Chap. 7).

This work provides evidence for a sharply localized electrical double layer, when strong correlations are present, confirming a common prediction of many ion correlation models. We anticipate that the results reported here, to be of relevance in other strongly correlated soft matter systems.

References

1. Holm, C., Kekicheff, P., Podgornik, R. (eds.): Electrostatic Effects in Soft Matter and Biophysics. NATO Science Series II: Mathematics, Physics and Chemistry. Kluwer Academic Publishers, Dordrecht (2000)
2. Poon, W.C.K., Andelman, D. (eds.): Soft Condensed Matter Physics in Molecular and Cell Biology. CRC Press; Taylor and Francis Group, Boca Raton (2006)
3. Levin, Y.: Electrostatic correlations: from plasma to biology. Rep. Prog. Phys. **65**, 1577 (2002)

4. Jacob, N.: Israelachvili. Intermolecular and Surface Forces. Academic Press, London (1992)
5. Gouy, G.: Constitution of the electric charge at the surface of an electrolyte. J. Phys. **9**, 457–467 (1910)
6. Chapman, D.L.: A contribution to the theory of electrocapillarity. Philos. Mag. Ser. 6, **25**, 475 (1913)
7. Stern, O.: Z. Elekt. Angew. Phys. Chem. **30**, 508 (1924)
8. Debye, P., Hückel, E.: The theory of electrolytes. I. lowering of freezing point and related phenomena. Physikalische Zeitschrift **24**, 185–206 (1923)
9. Antypov, D., Barbosa, M., Holm, C.: Incorporation of excluded-volume correlations into Poisson-Boltzmann theory. Phys. Rev. E **71**(6), 061106 (2005)
10. Bedzyk, M.J., Bommarito, G.M., Caffrey M., Penner, T.L.: Diffuse-double layer at a membrane-aqueous interface measured with x-ray standing waves. Science **248**, 52 (1990)
11. Park, C., Fenter, P.A., Sturchio, N.C., Regalbuto, J.R.: Probing outer-sphere adsorption of aqueous metal complexes at the oxide-water interface with resonant anomalous x-ray reflectivity. Phys. Rev. Lett. **94**, 076104 (2005)
12. Andresen, K., Das, R., Park, H.Y., Smith, H., Kwok, L.W., Lamb, J.S., Kirkland, E.J., Herschlag, D., Finkelstein, K.D., Pollack, L.: Counterion distribution around DNA probed by solution x-ray scattering. Phys. Rev. Lett. **93**, 248103 (2004)
13. Luo, G., Malkova, S., Yoon, J., Schultz, D.G., Lin, B., Meron, M., Benjamin, I., Vanysek, P., Schlossman, M.L.: Ion distributions at the nitrobenzene-water interface electrified by a common ion. J. Electroanal. Chem. **593**, 142–158 (2006)
14. Luo, G., Malkova, S., Yoon, J., Schultz, D.G., Lin, B., Meron, M., Benjamin, I., Vanysek, P., Schlossman, M.L.: Ion distributions near a liquid-liquid interface. Science **311**, 216–218 (2006)
15. Laanait, N., Yoon, J., Hou, B., Vanysek, P., Meron, M., Lin, B., Luo, G., Benjamin, I., Schlossman, M.: Communications: monovalent ion condensations at the electrified liquid/liquid interface. J. Chem. Phys. **132**, 171101 (2010)
16. Laanait, N., Mihaylov, M., Hou, B., Yu, H., Vanýsek, P., Meron, M., Lin, B., Benjamin, I., Schlossman, M.L.: Tuning ion correlations at an electrified soft interface. Proceedings of the National Academy of Sciences **109**(50), 20326–20331 (2012)
17. Torrie, G.M., Valleau, J.P.: Electrical double layers. I. Monte carlo study of a uniformly charged surface. J. Chem. Phys. **73**, 5807 (1980)
18. Patey, G.N.: The interaction of two spherical colloidal particles in electrolyte solution. an application of the hypernetted chain approximation. J. Chem. Phys. **72**(10), 5763 (1980)
19. Guldbrand, L., Jönsson, B., Wennerström, H., Linse, P.: Electrical double layer forces. a monte carlo study. J. Chem. Phys. **80**(5), 2221 (1984)
20. Derjaguin, B., Landau, L.: Theory of the stability of strongly charged lyophobic sols and of the adhesion of strongly charged particles in solutions of electrolytes. Acta Phys. Chem. URSS **14**, 633 (1941)
21. Verwey, E.J., Overbeek, J.T.G.: Theory of the Stability of Lyophobic Colloids. Elsevier, Amsterdam (1948)
22. Bloomfield, V.A.: DNA condensation. Curr. Opin. Struc. Biol. **6**(3), 334–341 (1996)

Chapter 2
The Poisson-Boltzmann Equation

In this chapter, we give a theoretical treatment of the Poisson-Boltzmann equation, and discuss the various approximations that enter into it, setting the stage for the various theoretical extensions that are experimentally investigated in subsequent chapters. We derive the PB equation from a density functional theory. In this formalism, a free energy functional $\mathcal{F}[n(\mathbf{r})]$ is postulated, where $n(\mathbf{r})$ is the inhomogeneous density, then the equations of motions are derived using a variational principle [1]. Once $\mathcal{F}[n(\mathbf{r})]$ is known, one can solve the resultant Euler-Lagrange equations (generally a system of differential equations) or use functional minimization, where successive self-consistent density profiles are tested to find the equilibrium (extremum) density profile [2]. Other than its mathematical simplicity, the DFT approach shifts the emphasis from particles and their interactions to densities. Hence, a large number of known equations of state of fluids can be used as density functionals, and their adequacy to describe the physics of a particular system easily determined. Moreover, including different types of correlations is done with ease, just by adding their contribution to the free energy. Needless to mention, that the difficulty in this approach resides in finding *some* functional that properly describes the system under study. Finally, the DFT formalism directly provides us with a free energy expression, from which various physical quantities (pressure, etc ...) are easily obtained to be compared with thermodynamic measurements. The Helmholtz free energy, $\mathcal{F}[n(\mathbf{r})]$ can be minimized subject to the constraint of particle conservation, or in a system with charged particles, the constraint of electroneutrality. Equivalently, one can use the grand potential, Ω, defined by $\Omega = \mathcal{F} - \mu N$, where μ is the chemical potential. Various subtleties arise in defining the grand potential as a unique functional of the density, instead of a generalized external potential. For instance, in a system with electrostatic interactions, the external potential is given by $\mu - e\phi(z)$, where $\phi(z)$ is the electrostatic potential. Nevertheless, once can prove that $\Omega[n]$ is a well-defined functional, uniquely determined by $n(\mathbf{r})$ (see [1]). Since, we are interested in studying a liquid/liquid interface under the application of an electric field normal to the interface, we assume homogeneity in the xy-plane, with the inhomogeneous density depending only on the z-axis, then the relevant grand potential for our system is

N. Laanait, *Ion Correlations at Electrified Soft Matter Interfaces*,
Springer Theses, DOI: 10.1007/978-3-319-00900-1_2,
© Springer International Publishing Switzerland 2013

$\Omega[n(z)]/A$, with A being the surface area. The Poisson-Boltzmann equation for a $v:v$ electrolyte that consists of $\pm$ ions in a medium with dielectric constant ϵ, can be derived using a density functional as follows,

$$\frac{\beta}{A}\Omega^{PB}[n(z)] = \sum_{i=+,-} \left(\frac{\beta}{A}\mathcal{F}^{PB}[n_i(z)] - \beta\int dz n_i(z)\mu_i\right) \quad (2.1)$$

where n_i is the 1-particle density of ion i, μ_i is the chemical potential. The free energy due to the ($\pm$) ions, $\mathcal{F}[n_\pm(z)]$ consists of the entropy in the ideal gas approximation, and the electrostatic interaction in the mean-field approximation, where the ions only interact with the average electrostatic potential in the system,

$$\frac{\beta}{A}\mathcal{F}^{PB}[n_\pm(z)] = \int n_\pm(z)\left(\ln(\Lambda^3 n_\pm(z)) - 1\right) dz \pm \frac{\beta}{2}ev\int n_\pm(z)\phi(z)dz \quad (2.2)$$

where Λ is the thermal wavelength. Varying the grand potential with respect to $n_\pm(z)$, we find

$$\frac{\beta}{A}\frac{\delta\Omega^{PB}[n(z)]}{\delta n_\pm(z')} = \int \left\{\frac{\delta n_\pm(z)}{\delta n_\pm(z')}(\ln(\Lambda^3 n_\pm(z)) - 1) + n_\pm(z)\left(\frac{\delta n_\pm(z)}{\delta n_\pm(z')}\frac{1}{n_\pm(z')}\right)\right\} dz$$
$$\pm \frac{\beta}{2}ev\left\{\int \frac{\delta n_\pm(z)}{\delta n_\pm(z')}\phi(z)dz + \int n_\pm(z)\frac{\delta\phi(z)}{\delta n_\pm(z')}dz\right\} - \beta\mu_\pm \int \frac{\delta n_\pm(z)}{\delta n_\pm(z')}dz$$
$$(2.3)$$

using the identity, $\frac{\delta n(z)}{\delta n(z')} = \delta(z - z')$, and assuming that $\phi(z)$ obeys Poisson's equation, and therefore can be expressed as,

$$\phi(z) = \int \mathcal{G}(z - z')(n_+(z') + n_-(z'))dz', \quad (2.4)$$

where $\mathcal{G}(z - z')$ is the Green's function. The functional derivative of the grand potential with respect to the density is

$$\frac{\beta}{A}\frac{\delta\Omega^{PB}[n_\pm]}{\delta n_\pm} = \ln(\Lambda^3 n_\pm) \pm \beta ve\phi(z) - \beta\mu_\pm$$

$$\Rightarrow n_\pm(z) = \frac{e^{\beta\mu_\pm}}{\Lambda^3}e^{\mp\beta ve\phi(z)} \quad (2.5)$$

In the 2nd line, we set the variation of the grand potential to zero and solved for $n_\pm$. Since, we only consider dilute solutions in this work, where the activity coefficient is very close to 1 (see [3]), then in the mean-field approximation, the chemical potential is given by $\mu_\pm = 1/\beta \ln(\Lambda^3 n^b)$, with n^b the ionic bulk density. From this expression, we obtain the Boltzmann distribution of the ions in the electrostatic potential,

2 The Poisson-Boltzmann Equation

$$n_{\pm}(z) = n^b e^{\mp \beta v e \phi(z)}, \tag{2.6}$$

This expression for the density distribution together with Poisson's equation gives the celebrated Poisson-Boltzmann equation (PB). In the application of the PB equation to a liquid/liquid interface, the Boltzmann distribution (Eq. 2.6) is defined in each phase from the corresponding solution concentrations defined in Chap. 3 from equilibrium partitioning. Then the PB equation is solved across the interface numerically subject to boundary conditions defined in Sect. 5.4. Typical density profiles from the PB equation are illustrated in Fig. 2.1. The key characteristic of these profiles is their monotonic decay away from the interface. Analytic density profiles can be derived by solving the PB equation in planar geometry when the surface charge or the potential $\phi(z)$ are defined at $z = 0$ (both of these cases do not apply to an electrified liquid/liquid interface). These exact ion distributions are found to follow an inverse power law behavior ($\sim 1/z^2$). In Chaps. 3, 4 and 6, we will show that the PB gives an inadequate description of the ion density profiles at electrified soft interfaces, both qualitatively and quantitatively. This inadequacy is easily traced back to a neglect of a myriad of interactions that an ion experiences in a physical system. In fact, as is clear from the DFT derivation above, the PB theory describes an *ideal* solution where the particles interact by a Coulomb potential. The neglect of ion-solvent, solvent-solvent, and ion-ion interactions in this mean-field model ultimately leads to contradictions with experimental results when the latter have sufficient resolution such as x-ray scattering techniques.

Heuristically, one can describe ion interactions "beyond" the mean-field approximation of the PB theory, by writing the ion distribution as follows,

$$n_i(z) \propto \exp\left(-\beta(\pm v_i e \phi(z) + W_i(z))\right) \tag{2.7}$$

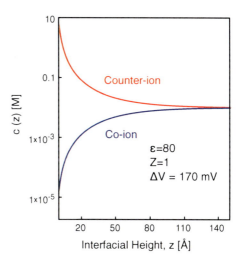

Fig. 2.1 Density profiles predicted by the poisson-boltzmann equation

where $W_i(z)$ is the potential of mean force due to all the solvent particles and ions in the system that interact with ion i, and that does not include the Coulomb interaction. In Chap. 4, we follow this approach to quantify ion interactions in *excess* of the ideal PB description of the electric double layer, by fitting $W_i(z)$ to the x-ray reflectivity data. In light of the comparison between PB predictions and the x-ray data, we find that PB ion density predictions "diverge" at highly polarized liquid/liquid interfaces, which motivates us to study steric effects in these systems (see Chap. 5) using a modified PB equation in the mean-field approximation, [3] based on using a lattice gas expression of the entropy instead of the ideal gas approximation used in (2.1). Restricting the ions to a lattice gives rise to repulsive short-range attractions, causing the density profiles to saturate in the close packing limit ($\sim 1/a_i^3$), where a_i is the effective ionic diameter, and are found to be given by

$$n_i(z) \propto \frac{\exp(-\beta(\pm v_i e\phi(z)))}{1 - 2a_i^3 n_b + 2a_i^3 n_b \cosh(ze\beta\phi(z))} \tag{2.8}$$

We find that this modified PB model may be of utility in describing the double layer near moderately charged soft interfaces. An accurate description of ion correlations both electrostatic and due to specific solvent correlations is obtained through a density functional formalism presented in Chap. 7, where the density profiles derived have the following form,

$$n_i(z) \propto \exp\left(-\beta(\pm v_i e\phi(z) + f_i^{sol}(z) + \mu_i^{ion}(z))\right) \tag{2.9}$$

where f_i^{sol} represents a free energy profile from the contributions of ion-solvent interactions that we simulate using Molecular Dynamics in Chap. 6, while μ_{ion} is the excess chemical potential due to electrostatic ion correlations that is calculated using a nonlocal density functional of the Debye-Hückel-Hole theory of a one-component plasma [4].

References

1. Hansen, J.-P., McDonald, I.: Theory of simple liquids, 3rd edn. Elsevier, Amsterdam, Netherlands (2006)
2. Deserno, M.: A Monte-Carlo approach to Poisson-Boltzmann like free-energy functionals. Physica A Stat. Mech. Appl. **278**(3–4), 405–413 (2000)
3. Borukhov, I., Andelman, D., Orland, H.: Steric effects in electrolytes: a modified poisson-boltzmann equation. Phys. Rev. Lett. **79**, 435 (1997)
4. Nordholm, S.: Simple analysis of the thermodynamic properties of the one-component plasma. Chem. Phys. Lett. **105**, 301 (1984)

Chapter 3
Electrochemical Methods

Historically, the study of electrical double layers originated in electrochemistry. The first double layer model (Gouy-Chapman theory) was motivated by the study of ions near an electrode in an electrochemical cell. Initially, electrochemistry was mainly focused on quantifying electrode processes, such as rates of reduction/oxidation, and the behavior of charge transport at the electrolyte/electrode interface. These early studies of electrochemical phenomena were restricted by the classical electrochemical measurements of current and potential, such as impedance spectroscopy (Sect. 3.3), and cyclic voltametry (Sect. 6.1). With the advent of modern electrochemical techniques, such as electroreflectance spectroscopy, electrochemists are able to study, in detail, the electronic structure of the electrolyte/electrode interface and its response to the electric field of the double layer, thereby complementing information gained through classical measurements.

However, direct measurements of the distribution of ions and solvent behavior at electrodes are out of the reach of current electrochemical methods. Although electrochemistry is invaluable in defining a number of important physical properties of the double layer, such as the potential of zero charge (PZC) and the differential capacitance, it cannot be used to gain insight into the energetics of the double layer such as ion adsorption and ion density profiles. For traditional electrochemical experiments, being of a thermodynamic nature, are devoid of structural information, and require additional non-thermodynamic assumptions about the microscopic behavior of the electrolytes to describe physical interactions of ions in solution. Often, those microscopic assumptions themselves are the subject of study in the structural measurements of x-ray reflectivity, yet electrochemical data used in conjunction with x-ray scattering studies, becomes an excellent tool to test the validity of models of ion distributions. (see Chaps. 5 and 7).

In the next sections we introduce the experimental systems under study and the electrochemical methods used to characterize the samples on which x-ray scattering studies were performed (see Chap. 4).

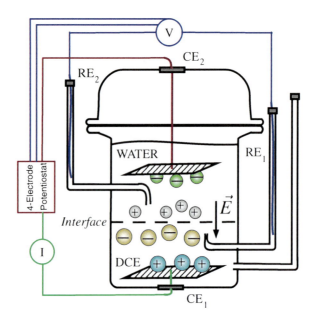

Fig. 3.1 Sketch of the electrochemical cell used in experiments of aqueous electrolyte/organic electrolyte. The cell diagram: Ag|AgCl|0.1 M NaCl + 20 mM HEPES (water) ||5 mM BTPPATPFB (DCE)|10 mM LiCl + 1 mM BTPPACl (water) |AgCl|Ag. The volume ratio of water:DCE is 2:1. An electric field at the interface causes the formation of a double layer on both sides of the interface

3.1 The Electrified Interface Formed by Two Immiscible Electrolyte Solutions

In what follows, we focus on the liquid/liquid interface formed by a 100 mM aqueous solution of NaCl including 20 mM HEPES to buffer the pH to 7.0, and a 5 mM solution of bis(triphenyl phosphoranylidene) ammonium tetrakis(pentafluorophenyl) borate (BTPPATPFB) in 1,2-dichloroethane (DCE), shown in Fig. 3.1. The crystal structure of BTPPATPFB is shown in Fig. 3.2 (see Appendix for atomic coordinates) and was obtained from x-ray powder diffraction measurements (Personal communication from Prof. Petr Vanysek, Northern Illinois University, USA). This system represents the staple of electrified liquid/liquid interfaces that we have investigated, with other samples varying only in the aqueous electrolyte composition. Furthermore, the electrochemical procedures employed are identical. The aqueous and organic solvents have negligible mutual solubility; the solubility of DCE in water is 0.158 %/mol [1], while the solubility of water in DCE is 1.24 %/mol [2]. When the two electrolytes are put into contact, the aqueous ions (Na^+, Cl^-) and the organic ions ($BTPPA^+$, $TPFB^-$) will partition across the interface, thereby slightly affecting the initial bulk concentrations in both the aqueous phase (w) and the organic phase (o). This process will take place until chemical equilibrium is reached. The

3.1 The Electrified Interface Formed by Two Immiscible Electrolyte Solutions

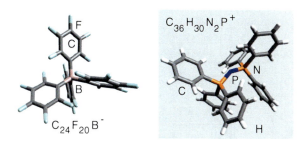

Fig. 3.2 Crystal structure of BTPPATPFB. Structure obtained from x-ray powder diffraction. *Left* TPFB$^-$, tetrakis(pentafluorophenyl) borate. *Right* BTPPA$^+$, bis(triphenyl phosphoranylidene) ammonium

initial and equilibrated concentrations of the ionic species i are given by $c_{i,0}^{w(o)}$ and $c_{i,eq}^{w(o)}$, respectively. Equilibrium between the two liquid phases necessitates the equality of the chemical potential for each ion, $\mu_i^o = \mu_i^w$, with μ_i defined as follows,

$$\mu_i = \mu_i^{id} + z_i F \phi$$
$$= \mu_i^{\circ} + RT \ln c_{i,eq} + z_i F \phi$$

where μ_i^{id} is the chemical potential of the ideal solution given in the second line, μ_i° is the standard chemical potential, z_i is the valency, and F is Faraday's constant. Using the ideal gas chemical potential amounts to setting the ionic activity, γ_i, to unity. In the case of the organic electrolyte, the concentration is low and BTPPATPFB is not fully dissociated (see Sect. 3.3). For the aqueous electrolyte, the activity deviates slightly from 1 [3], introducing negligible errors in the partitioned concentrations of Na$^+$ and Cl$^-$. Hence, we may set $\gamma_i = 1$ throughout the calculation. Using the equilibrium condition ($\mu_i^o = \mu_i^w$), gives the Nernst equation,

$$\Delta \phi_o^w = \frac{1}{z_i F}(-\Delta G_i^{\circ w \to o} + RT \ln \frac{c_{i,eq}^o}{c_{i,eq}^w}) \quad (3.1)$$

where $\Delta \phi_o^w = \phi^w - \phi^o$ is the *inner potential difference* between the two phases, $\Delta G_i^{\circ w \to o} = \mu_i^{\circ w} - \mu_i^{\circ o}$ is the standard Gibbs energy of transfer of ion i from the water phase to the oil phase (DCE). This change in free energy due to ionic partitioning is a measurable quantity. The energies of transfer of the aqueous ions, Na$^+$ and Cl$^-$, are widely available in the literature, for instance, see [4] and references within. The Gibbs energy of transfer of the organic ions (TPFB$^-$, BTPPA$^+$) were measured by UV-visible spectroscopy and mass spectroscopy [5], and shown in Table 3.1.

In contrast to $\Delta G^{\circ w \to o}$, the inner potential difference is not a measurable quantity but needs to be solved for as well. Hence, if the number of species of ions is n, then we need to determine $2n$ concentrations ($c_{i,eq}^w, c_{i,eq}^o$) producing $2n+1$ unknowns. In addition, we have n Nernst equations, n constraints given by the initial concentrations,

Table 3.1 Free energies of transfer and equilibrated ionic bulk concentrations

	$\Delta G_i^{\circ \, w \to o}$ (kJ/mol)	$c_{i,0}^w$ (mM)	$c_{i,0}^o$ (mM)	$c_{i,eq}^w$ (mM)	$c_{i,eq}^o$ (mM)
Na$^+$	57 ± 6[a]	100.0	0.0	100.0	1.4×10^{-8}
Cl$^-$	53 ± 4	100.0	0.0	100.0	3.7×10^{-8}
BTPPA$^+$	−56 ± 2	0.0	2.689	2.9×10^{-10}	2.689
TPFB$^-$	−72.5 ± 6	0.0	2.689	7.4×10^{-13}	2.689

[a] Values of aqueous ions from [4]

and a bulk electroneutrality condition in either phase,

$$\sum_i z_i c_{i,eq}^w = 0,$$

producing a solvable system of equations. The equilibrated bulk concentrations are shown in Table 3.1. The bulk ionic concentration of BTPPA$^+$ and TPFB$^-$ is different than the concentration of BTPPATPFB (=5 mM), primarily as a results of partial dissociation, as determined in Sect. 3.3. From Table 3.1, we remark that the final equilibrated concentrations are practically the same as the initial concentrations, allowing us to ignore partitioning, when convenient. While, the above treatment is exactly valid in the absence of external fields, it is approximately valid when an external electrostatic potential is applied, and care must be taken when applying large potentials, as discussed below.

The solutions were prepared from purified solvents. Water was produced by a Barnstead Nanopure system and DCE was purified by multiple passings through a column of basic alumina. NaCl in the form of powder was purchased from Fisher Scientific and further purified by roasting to remove water and impurities. BTPPATPFB was synthesized from BTPPACl (Aldrich) and LiTPFB (Boulder Scientific) as described in [5, 6]. To allow for the two electrolytes to saturate and reach equilibrium, the two phases are put in contact in a beaker and rocked for ten hours; this allows for diffusion processes and ion partitioning to take place. The solutions are then extracted separately and put in the sample cell, see Fig. 3.1. We found that saturation of the phases is crucial to producing a structurally stable liquid/liquid interface when probed by x-ray scattering.

The electrolyte solutions are hosted in a glass electrochemical cell shown in Fig. 3.1, where the liquid/liquid interface has a diameter of 7 cm. An electric field is applied at the interface using the four-electrode potentiostat method, where a current is forced through the counter electrodes (CE$_{1,2}$) made from a platinum mesh with area 9 cm^2 and the potential is monitored at the reference electrodes (RE$_{1,2}$). The desired applied potential is obtained by the potientiostat through a feedback loop, between the current collected at CE$_{1,2}$ and the potential at RE$_{1,2}$. Ag/AgCl reference electrodes are inserted into Luggin capillaries that terminate within 4 mm of the interface. Being in close proximity to the interface and with negligible current passing through them (i.e. in chemical equilibrium), the reference electrodes enable very accurate and

3.1 The Electrified Interface Formed by Two Immiscible Electrolyte Solutions

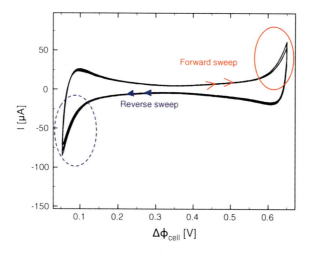

Fig. 3.3 Cyclic Voltammetry plot for the 100 mM NaCl (water)/5 mM BTPPATPFB (DCE) system. The potential is scanned at a rate of 10 mV/s, while collecting the current. The *circles* indicate the onset of ion transfer across the interface (see text). The measurement reveals a polarization window in the range of $\Delta\phi_{cell}^{w\text{-}o} \approx 0.05\text{--}0.65$ V

stable monitoring of the applied potential at the interface. The Luggin capillary on the DCE side contains a solution of 10 mM LiCl + 1 mM BTPPACl, with LiCl added to compensate for the low ionic conductivity of the organic salt BTPPACl. The electric field at the water/DCE interface causes the ions to form back-to-back double layers at the interface. The structure of the double layer is controlled by tuning the amplitude of the electric field and is then probed by x-ray reflectivity measurements (see Chap. 4).

We conduct cyclic voltammetry measurements (CV) to find the system's polarization window. This determines the range of interfacial potentials where the interface is ideally polarizable, i.e the system is at equilibrium in the absence of irreversible transfer of ions across the interface. Cyclic voltammetry is the most widely used electrochemical technique and can be employed to measure ion diffusion coefficients, free energies of transfer and a host of other properties [7]. CV measurements are performed by linearly sweeping the applied potential in the forward direction shown by open arrows in Fig. 3.3), $\Delta\phi_{cell}^{w\text{-}o} = \phi^{RE_2} - \phi^{RE_1}$ at a scan rate of 10 mV/s, up to some maximum potential value $\Delta\phi_{cell}^{w\text{-}o} = 0.65$ V, while recording the current. Then, reversing the sweep (closed arrows in Fig. 3.3) and reducing the potential to a minimum value $\Delta\phi_{cell}^{w\text{-}o} = 0.05$ V, and then back to its starting value. This is done for a few cycles to ensure reproducibility of the sample's response to the electric field, and as a test for cleanliness, as shown in Fig. 3.3. The extrema of $\Delta\phi_{cell}^{w\text{-}o}$ are carefully chosen to be bounded by the Gibbs energies of transfer of the ions (Table 3.1) in the system to avoid transferring ions across the interface. In fact, if the scan was extended further, a well-defined and symmetric current peaks would show up in the CV curve at a potential value given by $\frac{1}{zF}\Delta G^{w\rightarrow o}$. Note that we have omitted the standard symbol, since a reference state needs to be chosen, (requiring an extra-thermodynamic assumption) before this potential can be interpreted as a standard free energy of transfer [7]. Also a knowledge of the $\Delta G_i^{\circ\, w \to o}$ for the other electrolyte is necessary.

As seen in Fig. 3.3, for $\Delta \phi_{cell}^{w-o} > 0.5$ V, a positive current starts to flow due to double layer charging, that is enhancement of TPFB$^-$ on the DCE side and Na$^+$ on the water side. There is also a contribution of the current from a small amount of ion transfer across the interface. As one approaches $\Delta \phi_{cell}^{w-o} \approx 0.65$ V, there is a significant increase in the positive current indicating the onset of ion transfer (solid circle in Fig. 3.3), that is TPFB$^-$ to the water side and Na$^+$ to DCE side. This indicates that potentials higher than 0.65 V should not be applied in order to avoid large current flow. In fact, the magnitude of the current is proportional to the sweep rate, which is greatly reduced during the reflectivity measurements compared to that used during CV, leading to a system in a steady-state of equilibrium with minimal current ($\ll 5$ µA). For instance, after applying a potential (within the polarization window) the current is high initially, but then drops and reaches a steady-state value, as shown in Fig. 3.4. Similarly, we determined that potentials lower than 0.05 V should not be applied due to transfer of BTPPA$^+$ to the water phase and Cl$^-$ to the DCE phase (dashed circle in Fig. 3.3). The polarization window is roughly estimated from a CV measurement. During x-ray experiments, the system is subject to an applied potential for much longer times than during CV measurements, on the order of minutes or hours. Hence, the polarization window obtained from CV is further verified by checking the reproducibility of the x-ray intensity as a function of $\Delta \phi^{w-o}$ and as a function of time as discussed in Chap. 4.

The cyclic voltammetry curve of a reversible and stable system should exhibit the following properties, discussed in [8]:

- The current peak position does not change as a function of potential sweep rate.

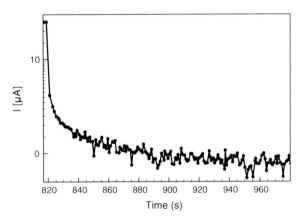

Fig. 3.4 Plot of the current as a function of time during the application of an electrostatic potential

3.1 The Electrified Interface Formed by Two Immiscible Electrolyte Solutions

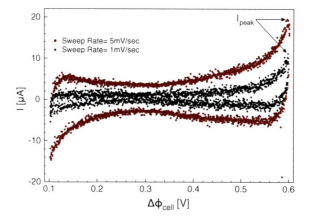

Fig. 3.5 Varying the potential sweep rate of the CV data for the 100 mM NaCl (water)/5 mM BTPPATPFB (DCE) system. The potential is scanned at two different rates of 5 and 1 mV/s. The plots exhibit the properties of a stable, reversible system

- The peak current magnitude, I_1 and I_2, taken at two different potential sweep rates, ν_1 and ν_2, should obey:

$$\frac{I_1}{I_2} = \sqrt{\frac{\nu_1}{\nu_2}}$$

This equation can be derived by assuming that $c_i^{w(o)}$ obey the Fick diffusion equation,

$$D_i^{w(o)} \frac{\partial^2 c_i^{w(o)}(z,t)}{\partial z^2} = \frac{\partial c_i^{w(o)}(z,t)}{\partial t},$$

where $D_i^{w(o)}$ represents the diffusion constant of ion i, z is the interfacial normal and t is the time variable. The boundary conditions are consistently determined by Eq. (3.1).

In Fig. 3.5, the CV data of the 100 mM NaCl (H$_2$O)/5 mM BTPPATPFB (DCE) is shown, using two different potential sweep rates, 5 and 1 mV/s. As mentioned the current peak position is independent of the sweep rate. Furthermore, $\frac{I_1}{I_2} = 1.74$ is in fair agreement with $\sqrt{\frac{\nu_1}{\nu_2}} = 2.24$, which indicates that the overall electrolyte/electrolyte system is stable and reversible.

3.2 The Potential of Zero Charge

The interfacial potential $\Delta\phi^{\text{w-o}}$ is distinct from the applied potential $\Delta\phi^{\text{w-o}}_{\text{cell}}$. Since, we expect that at $\Delta\phi^{\text{w-o}} = 0\,\text{V}$, minimal current should be present (i.e. no surface excess charge on either side of the interface), this is seen by close inspection of Fig. 3.3 to be the case when $\Delta\phi^{\text{w-o}} \approx 0.35\,\text{V}$. Hence, a determination of the "zero" of the interfacial potential is needed. We define the interfacial potential as $\Delta\phi^{\text{w-o}} = \Delta\phi^{\text{w-o}}_{\text{cell}} - \Delta\phi^{\text{w-o}}_{\text{PZC}}$, where the last term is called the potential of zero charge (PZC). In subsequent chapters, we abbreviate our notation, denoting the interfacial potential by $\Delta\phi$ only. The fact that the value of this potential is nonzero, should not come as a surprise, since the potential we measure, $\Delta\phi^{\text{w-o}}_{\text{cell}}$, is across the entire electrochemical cell, not just the liquid/liquid interface. Hence, to properly account for the electrostatic potential that the ions "feel" at the interface when polarized, we need to determine a value of $\Delta\phi^{\text{w-o}}_{\text{PZC}}$. This will be eventually used in our description of the ions' energetics. One method of finding the PZC is accomplished by its thermodynamic relation to the interfacial tension, a property that is easily measurable. Before we present these measurements, a brief digression into the thermodynamics of surfaces is necessary.

Since thermodynamic quantities are defined and determined in the bulk without structural assumptions, a strict application of the thermodynamic methodology necessitates that quantities related to the surface or interface are to be defined as *excess* values over those of the bulk. This is the approach originated by Gibbs [9, 10].

Consider two semi-infinite phases α and β in contact over a planar area A, as in Fig. 3.6. For instance, α and β could represent two immiscible liquids or a liquid phase in contact with a gas phase. In the canonical ensemble, the bulk phases are completely

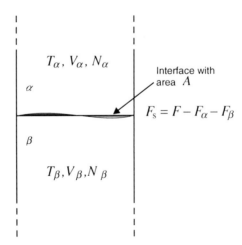

Fig. 3.6 Illustration of the thermodynamic procedure to define surface excess properties

3.2 The Potential of Zero Charge

determined by their values of temperature, volume, and number of particles. We define the total volume of the system by $V = V_\alpha + V_\beta$ this choice depends on a choice of a mathematical surface, termed the *Gibbs dividing surface* (GDS). However, this choice is arbitrary as any surface with the "correct" normal vector would do. Hence, we pick a GDS so that $\sum_i \mu^i N_s^i = 0$, where N_s^i is the number of particles at the interface. The appropriate thermodynamic function is the Helmholtz free energy F, given below for each bulk phase:

$$F_\alpha = -pV_\alpha + \sum_i \mu^i N_\alpha^i, \qquad F_\beta = -pV_\beta + \sum_i \mu^i N_\beta^i \qquad (3.2)$$

i is summed over all particle species present in the bulk phases. The total free energy F of the system cannot simply be the sum of the bulk contributions:

$$F_\alpha + F_\beta = -pV + \sum_i \mu^i N^i. \qquad (3.3)$$

since the quantity in (3.3) does not depend on all the extensive variables of the system: V, N, and A. This excess free energy $F_s(A)$, is interpreted as the free energy of the interface. The total free energy of the system is given by:

$$F = F_\alpha + F_\beta + F_s(A) \qquad (3.4)$$
$$= -pV + \sum_i \mu^i N^i + \gamma A \qquad (3.5)$$

where γ is the energy cost to create a surface with area A under the conditions of isothermal, reversible work. We identify it with the interfacial tension. Numerous methods exist to measure this work. We employ the Wilhelmy plate method [11], where a plate is put in wetting contact with the interface. The interfacial tension, then is the force per length required to pull the plate out of the liquid, (note that this force only has a *tangential* component, as needed to properly identify it with the interfacial tension "force"). The teflon plate is attached to a Cahn microbalance, that measures the force per length.

The surface free energy per unit area at an interface formed between two electrolyte solutions, as a result of charging of the interface, is

$$dF = \gamma + \sigma d(\Delta \phi_{\text{cell}}^{\text{w-o}}) \qquad (3.6)$$

where σ is the surface excess charge per unit area. At equilibrium (constant temperature and chemical equilibrium) $\partial F/\partial \Delta \phi_{\text{cell}}^{\text{w-o}} = 0$, and we have the Lippmann equation:

$$\sigma = -\left(\frac{\partial \gamma}{\partial \Delta \phi_{\text{cell}}^{\text{w-o}}}\right)_{T,V,\mu}. \qquad (3.7)$$

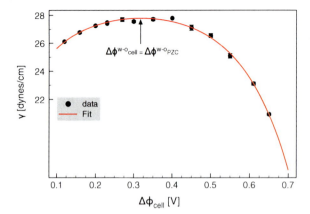

Fig. 3.7 Potential-dependent tension data of the 100 mM NaCl (aqueous)/5 mM BTPPATPFB system. Tension measurements taken using the Wilhelmy plate method, as a function of $\Delta\phi_{cell}^{w-o}$ to determine the potential of zero charge

Due to the large cross-sectional area of the cell, and those of the electrodes, small current densities are generated, smaller than a few tens of μA/cm^2. These current densities, generate insignificant Joule heating, justifying the constant temperature assumption. Furthermore, prior saturation of the two liquid phases as discussed earlier, guarantees that the chemical potential, most importantly μ^{ions}, is constant.

The Lippmann equation allows us to determine the surface excess charge from potential-dependent tension measurements. Figure 3.7 shows tension data taken on the 100 mM NaCl (aqueous)/5 mM BTPPATPFB system as a function of applied potential, which is equal to the *potential of zero charge*, when $\gamma(\phi)$ is maximum. The data is fitted against a polynomial of even powers of ϕ (up to ϕ^4), producing $\Delta\phi_{PZC}^{w-o} = 318 \pm 3$ mV. The latter is in good agreement with our estimate from the CV measurements.

In addition to determining the PZC, measurements of the surface excess charge provide an excellent tool to probe the accuracy of theories of the double layer. However, before we undertake such studies, we need to address the bulk behavior of ions.

3.3 Conductivity Measurements of the Dissociation of BTPPATPFB

Due to the low permittivity of 1,2-Dichloroethane ($\varepsilon = 10.43$), it is expected that BTPPATPFB would not fully dissociate into its ionic constituents. A rudimentary explanation of this fact is easily given in terms of a crude solvation picture. Suppose that we have an electrolyte with fully dissociated monovalent ions (e.g. Na$^+$, Cl$^-$)

3.3 Conductivity Measurements of the Dissociation of BTPPATPFB

in a polar solvent such as water ($\varepsilon = 78.95$) with a Bjerrum length $l_B \approx 7$ Å, at room temperature. Henceforth, for a separation $r \gtrsim 7$ Å, the ions Na$^+$ and Cl$^-$ are quite stable as a dissociated compound due to thermal fluctuations in the electrolyte. Furthermore, the ion–dipole interaction energy ($\mu e/r^2 \varepsilon$) for a highly polar molecule such as water ($\mu = 1.85$ D) at molecular separation, is on the order of kT; leading to a further stabilization of the ion by solvation. DCE with a $l_B \approx 55$ Å, would cause BTPPA$^+$ and TPFB$^-$ to be unstable due to the strength of their Coulomb attraction and the negligible ion–solvent dipole interaction. Thereby, a significant fraction would associate into the compound BTPPATPFB. Evidently, this crude picture omits a number of important physical effects such as the dependence of screening on the ionic concentration and specific solute–solvent interactions to be of any quantitative use. However, the above considerations demonstrate the necessity to measure the degree of dissociation of BTPPATPFB in 1,2-Dichloroethane.

We determine the degree of dissociation by performing solution conductivity measurements. Although, modern techniques such as IR Raman spectroscopy [12] or NMR [13] are nowadays routinely employed to measure the dissociation constant in a number of biological and chemical settings, where the complexity of the latter, makes the interpretation of the conductivity data difficult. For the purpose of measuring the dissociation of electrolytes, conductivity measurements remain an accurate method, simple in both its data interpretation and experimental setup.

The experimental apparatus consists of an electrochemical cell resembling the one portrayed in Fig. 3.8 with solid platinum plates immersed in an organic solution

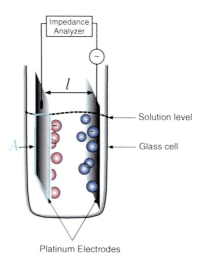

Fig. 3.8 Illustration of the experimental setup of solution conductivity measuruments (not drawn to scale). A glass cell hosts an organic solution of BTPPATPFB, in contact with two solid electrodes separated by a distance l. A is the area of the electrodes exposed to the solution. Under bias, the ions migrate to the electrodes inducing a current. The conductivity is determined from the measured impedance

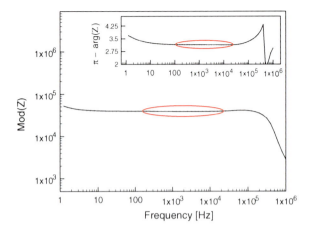

Fig. 3.9 Extracting solution Resistance from Impedance data. The *oval box* in the *inset* shows the range of frequencies over which the current and the potential are in phase, corresponding to $|Z|$ equivalent to solution resistance

of BTPPATPFB at some concentration c. The electrodes were rinsed with deionized water prior to each measurement and the cell was kept in a thermostatic bath. An AC voltage on the order of 10 mV is applied at the electrodes, resulting in an electric current due to transfer of charged ionic species. The response of the system is collected and processed by an impedance analyzer (Solartron 1255+1286). A large frequency range was used (1 Hz–1000 kHz), obtaining a plot of $|Z|$ versus Frequency as shown in Fig. 3.9, where Z is the impedance. When the phase shift θ between V and I is close to zero, $|Z|$ is essentially the cell resistance R, assuming Ohmic behavior. We determine R from the data by averaging over $|Z|$ in the relevant frequency range.

The electrolyte solution was prepared from purified DCE that was equilibrated with 100 mM NaCl aqueous solution as described in Sect. 6.1. A stock organic solution at a concentration of 100 mM is initially prepared from which solutions at other concentrations were obtained by dilution, see Table 3.2. The impedance was measured at each of these concentrations, from which the solution resistance is deduced as discussed. The solution conductance G, is the reciprocal of the resistance. To find the conductivity of the solution σ, the cell geometry as shown in Fig. 3.8, needs to be accounted for:

$$\sigma = G \frac{l}{A},$$

where l is the distance between the two electrodes and A is their cross-sectional area in contact with the solution. l/A is called the cell constant and is not easily determined from an accurate measurement of the cell's dimensions. Customarily, the cell constant is found through calibration. In our experiment, l/A is determined by calibration against a standard 0.01 M aqueous solution of potassium chloride, whose conductivity is known (1.413 mS/cm). Measuring the calibration solution conductance yields a cell constant of 8.81×10^{-3} cm^{-1}. The experimental quantity

3.3 Conductivity Measurements of the Dissociation of BTPPATPFB

Table 3.2 degree of dissociation of BTPPATPFB at various concentrations in 1,2-dichloroethane from solution conductivity measurements

c, mol/cm^3	G, Siemens ($\pm 4 \times 10^{-8}$)[a]	Λ, S · cm^2/mol ($\pm 2 \times 10^{-2}$)	θ, % (± 1)
2.00×10^{-9}	6.35×10^{-7}	2.80	99.15
5.00×10^{-9}	1.59×10^{-6}	2.79	98.96
5.00×10^{-8}	1.50×10^{-5}	2.64	94.50
1.00×10^{-7}	3.02×10^{-5}	2.66	95.77
2.00×10^{-7}	5.82×10^{-5}	2.56	92.86
5.00×10^{-7}	1.28×10^{-4}	2.25	82.68
1.00×10^{-6}	2.01×10^{-4}	1.78	66.09
2.00×10^{-6}	3.65×10^{-4}	1.61	60.82
5.00×10^{-6}	7.85×10^{-4}	1.38	53.79
1.00×10^{-5}	1.44×10^{-3}	1.26	50.84
2.00×10^{-5}	2.58×10^{-3}	1.14	48.16
5.00×10^{-5}	5.63×10^{-3}	0.99	45.51
1.00×10^{-4}	9.28×10^{-3}	0.82	40.64

[a]Only the largest error is quoted. The source of the uncertainty are fluctuations in the measured impedance

that is accessible theoretically is the equivalent conductance, $\Lambda = \dfrac{\sigma}{c}$, when defined in units of S · cm^2 · mol^{-1}.

Deriving an expression of Λ from a theory of electrolytes, that agrees with experimental data, was one of the most studied problems in the early years of investigating electrolytes, with contributions from a number of illustrious physicists including Debye and Onsager. To properly describe electrolytic conductance, one needs to consider not only the electrostatic interaction but also hydrodynamics due to the mobility of the charges as well as local gradients in the osmotic pressure and viscosity. For a thorough treatment (and a compelling historical perspective), the interested reader is referred to the book by Fuoss [14], one of the early contributors to the subject. An expression for the equivalent conductance is needed to interpret the data. We use the expression due to Fuoss and Onsager [15], which improved over the Debye-Huckel conductance by considering the above mentioned effects, in addition to electrostatic interactions

$$\Lambda = \Lambda_0 - S\sqrt{c} + E\,c\,\log c + J\,c \tag{3.8}$$

where Λ_0 is the limiting equivalent conductance, S and E are constants that depend on Λ_0, the solvent's permittivity, and viscosity. While J is also a constant, a function of the ion size, which are modeled as spheres.

Equation 3.8 is fitted against the data as shown in Fig. 3.10. Reasonable agreement is obtained over the entire range of concentrations. In the limit of infinite dilution, the Fuoss-Onsager equation does not seem to follow the data trend, giving $\Lambda_0 = 2.93 \pm 0.03$ S · cm^2/mol. Since, we are only interested in Λ_0 to calculate the

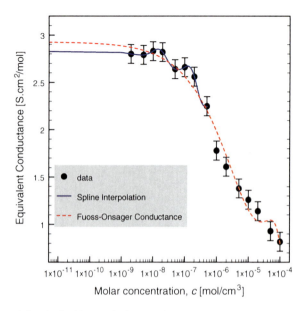

Fig. 3.10 Determining the limiting equivalent conductance

degree of dissociation at a specific concentration $c = 5$ mM, an accurate Λ_0 is needed. Due to the somewhat monotonic trend of Λ in the range of $c \leq 10^{-8}$mol · cm^{-3}, we expect that the limiting conductance could be extrapolated from measured values. We performed a spline interpolation on the data, shown in Fig. 3.10, producing $\Lambda_0 = 2.83$ S · cm^2/mol. Given the latter, the degree of dissociation could be determined from the Shedlovsky equation [16],

$$\Lambda = \theta \Lambda_0 - \alpha (\Lambda/\Lambda_0)(c\,\theta)^{1/2}, \tag{3.9}$$

$$\alpha = 8.2 \times 10^5 \Lambda_0/(\varepsilon\,T)^{3/2} + 82/\eta(\varepsilon\,T)^{1/2}$$

where α is the Onsager coefficient, η is the viscosity of DCE (=8.87 × 10^{-3} Pa · s), ε is the permittivity of DCE (=10.43), and T is the absolute temperature. The degree of dissociation of BTPPATPFB in 1,2-Dichloroethane is calculated from (3.9) and tabulated in Table 3.2. Note that using $\Lambda_0 = 2.93 \pm 0.03$ S · cm^2/mol gives a dissociation that is less than 1 % different than the values listed in the table.

We conclude that 53.79 % (±1 %, the uncertainty arises from fluctuations in the measured impedance) of BTPPATPFB is dissociated in the system we study, resulting in an ionic bulk concentration about half that of the initial bulk BTPPATPFB concentration. This charge carrier concentration is the meaningful quantity that enters electrostatic models used to predict the ion density profiles.

References

1. Weast, R.C. (ed.): CRC Handbook of Chemistry and Physics, 70th edn. CRC Press, Boca Raton (1989)
2. Matsuoka, I., Naito, T., Yamada, H.: Relationship between the solubility of water in organic solvents and the extraction equilibrium. Bunseki Kagaku **51**(9), 759–765 (2002)
3. Cohen, P. (ed.): The ASME Handbook on Water Technology for Thermal Power Systems. American Society of Mechanical Engineers, New York (1988)
4. Rais, J., Okada, T., Alexova, J.: Gibbs energies of transfer of alkali metal cations between mutually saturated water–solvent systems determined from extraction experiments with radiotracer 137 cs. J. Phys. Chem. B **110**, 8432–8440 (2006)
5. Hou, B.: Ion distributions at electrified liquid–liquid interfaces: an agreement between X-ray reflectivity analysis and macroscopic measurements. Doctoral dissertation, University of Illinois at Chicago (2011)
6. Fermin, D.J., Ding, Z., Duong, H., Girault, H.H.: Photoinduced electron transfer at liquid/liquid interfaces—part I: photocurrent measurements associated with heterogeneous quenching of zinc porphyrins. J. Phys. Chem. B **102**, 10334 (1998)
7. Bard, A.J., Faulkner, L.R.: Electrochemical Methods: Fundamentals and Applications, 2nd edn. Wiley, New York (2001)
8. Girault, H.H.: Analytical and Physical Electrochemistry. EPFL Press, Lausanne (2004)
9. Gibbs, J.W.: The Scientific Papers of J. Willard Gibbs, vol. 1: Thermodynamics. Ox Bow Press, Woodbridge (1993)
10. Rowlinson, J.S., Widom, B.: Molecular Theory of Capillarity. Clarendon Press, Oxford (1982)
11. Adamson, A.W.: Physical Chemistry of Surfaces, 5th edn. Wiley, New York (1990)
12. Edwards, H.G.M., Brown, D.R., Dale, J.R., Plant, S.: Raman spectroscopic studies of acid dissociation in sulfonated polystyrene resins. J. Mol. Struct. **595**(1–3), 111–125 (2001)
13. Kamatari, Y.O., Yokoyama, S., Tachibana, H., Akasaka, K.: Pressure-jump NMR study of dissociation and association of amyloid protofibrils. J. Mol. Biol. **349**(5), 916–921 (2005)
14. Raymond, F.M., Accascina, F.: Electrolytic Conductance. Interscience, New York (1959)
15. Fuoss, R.M., Onsager, L.: Conductance of unassociated electrolytes. J. Phys. Chem. **61**(5), 668–682 (1957)
16. Fuoss, R.M., Shedlovsky, T.: Extrapolation of conductance data for weak electrolytes. J. Am. Chem. Soc. **71**(4), 1496–1498 (1949)

Chapter 4
X-ray Reflectivity Studies of Ion Condensation at the Electrified Liquid/Liquid Interface

X-ray reflectivity (XR) is a powerful tool to probe the atomic structure of surfaces and interfaces. Due to the weak interaction of x-rays with matter, XR measurements are most suitable to study the interfacial structure of systems in their natural settings, i.e. in situ, specifically in extreme physical or chemical conditions, and when the interface is inaccessible to microscopy, as in buried interfaces. The study of liquid surfaces by synchrotron x-ray scattering was pioneered by Pershan, Als-Nielsen and coworkers in the early eighties [1], and was put on a firm theoretical footing with significant contribution from Pershan's group [2–4], as well as S. Sinha and coworkers [5], leading to the understanding that x-ray reflectivity probes gradients in the electron density in conjunction with the interface's long wavelength fluctuations. For further details on the subject, we refer the reader to the numerous reviews of XR applied to liquid surfaces [6, 7]. The first XR studies of water/oil interfaces were performed by M. Schlossman and coworkers, specifically water/alkane interfaces [8], where it was found that the structure of the interface is approximated by capillary wave theory [9], though an intrinsic interfacial structure that varies with the type of organic solvent is needed to account for the observed interfacial width (for a review see [10, 11]). Subsequent XR studies of water/polar oil interfaces focused on ion distributions near the water/nitrobenzene interface, where the interfacial potential was controlled by partitioning of a ion common to both phases [12, 13]. These studies showed that the electrical double layer energetics is greatly influenced by the interfacial liquid structure. Recently, we have shown using XR that ionic condensation takes place at a liquid/liquid interface under the application of an electric field [14]. These results are the focus of this chapter.

In Sect. 4.1, we briefly discuss experimental details pertinent to measurements of reflectivity from the electrified liquid/liquid interface. Procedures to analyze the x-ray data are presented in Sect. 4.2. We conclude with some remarks.

4.1 Experimental Procedure

X-rays interact very weakly with matter by virtue of the Thomson cross-section ($\sigma_s \approx 10^{-9} \text{Å}^2$ per electron). This is further accentuated at liquid interfaces where the number density is small, resulting typically in scattering events of 1 photon per 10^{10} incident photons. This weak interaction requires the use of very strong x-ray sources such as synchrotron storage rings. Hence, all of the x-ray experiments reported in this work have been done at the Advanced Photon Source (Argonne National Laboratory), a third generation synchrotron. Due to this weak scattering, x-ray reflectivity can be treated in a kinematical approximation where multiple scattering events are neglected. This is an excellent approximation for XR from liquid surfaces when conditions are far from the critical angle of total external reflection. In elastic scattering, the most relevant variable is the momentum transfer variable, **Q**, defined by

$$\mathbf{Q} = \mathbf{k}_{out} - \mathbf{k}_{in}$$

where $\mathbf{k}_{out}$ is the wavevector of the outgoing beam and $\mathbf{k}_{in}$ is the wavevector of the incident beam. If we are only interested in reflection in the specular condition, i.e the incident angle α is equal to outgoing angle, then the only non-vanishing component of the scattering vector **Q** is the component normal to the interface, as shown in Fig. 4.1, and given by

$$Q_z = \frac{4\pi}{\lambda} \sin \alpha$$

where λ is the beam's wavelength. The wavelength is related to the energy by hc/E, where h is Planck's constant and c is the speed of light. The energy of the x-ray beam

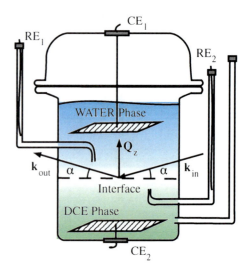

Fig. 4.1 X-ray kinematics and sample cell

4.1 Experimental Procedure

we use is determined by the characteristics of the sample studied. In our studies, as illustrated in Fig. 4.1, the x-ray beam needs to travel through a thick phase of water before reaching the interface, hence attenuation in the incident beam's intensity is unavoidable due to absorption. The absorption cross section is related to the energy through, $\sigma_a \propto 1/E^3$, which is valid far from an absorption edge. Therefore, it is beneficial to have as high an energy as possible. The beamline where experiments were performed, Sector 15 (ChemMatCars, University of Chicago), permitted us to reach a beam energy of 30 keV while keeping a high-flux, well-focused beam. All specular reflectivity data presented in this work was then performed at a wavelength $\lambda = 0.41255 \pm 0.00005$ Å.

In what follows, we consider an ideal flat surface (shown in Fig. 4.2), characterized by a step-function electron density profile $\rho(z)$, and study the dependence of its reflectivity on the momentum transfer Q_z, and the critical momentum transfer Q_c. For x-rays, $Q_c = 4\sqrt{\pi \Delta \rho r_e}$, where $\Delta \rho = \rho_2 - \rho_1$ the difference in electron densities of the bulk phases, and r_e is the Thomson scattering length, neglecting anomalous dispersion corrections. In the region of $Q_z \gg Q_c$, the reflectivity is given by the Fresnel reflectivity from classical electromagnetic theory,

$$R_F(Q_z) \sim \left(\frac{Q_c}{2Q_z}\right)^4$$

Since $R_F(Q_z)$ falls fast with Q_z, this limits the spatial resolution Δ accessible in a reflectivity measurement [15]. For instance, the water/DCE system has $Q_c \approx 8 \times 10^{-3}$ Å^{-1}, so to resolve molecular structure ($\Delta = 1$ nm), XR data needs to be measured up to $Q_{zmax} \approx \pi/\Delta = 0.3$ Å^{-1}. When $Q_z > 0.3$ Å^{-1}, bulk scattering overwhelms the specular signal and the reflected intensity drops to less than 10^{-11}.

Liquid interfaces are not ideal, with an electron density profile that varies smoothly with depth. Hence the reflectivity from the latter is not given by Fresnel reflectivity, but in the kinematic approximation (still assuming a flat surface) can be expressed as,

$$R(Q_z)/R_F(Q_z) \approx \frac{1}{\Delta \rho}|\hat{F}\{d\rho(z)/dz, Q_z\}|^2$$

Fig. 4.2 Illustration of an ideal interface

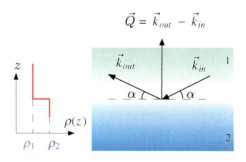

where $\hat{F}\{.,.\}$ represents the Fourier transform. This shows that XR is very sensitive to the interfacial normal *inhomogeneous* density profile of the electrical double layer. And this sensitivity can be tuned by progressive polarization of the interface, which leads to variations in the gradient of the electron density. However, XR only gives information that is averaged over the xy-plane and must be supplemented with in-plane scattering to probe $2d$ ordering. We note that we have been unsuccessful in our attempts to measure in-plane scattering at these electrified liquid/liquid interface, as a result of the strong scattering from the bulk liquid phases.

Measurements of the reflectivity were done using a liquid surface reflectometer at sector 15 [16]. The reflectometer consists mainly of a steering crystal to bring the beam to the sample, ion chambers to monitor the beam intensity, slits to define the beam cross-section and a sample stage where the sample resides. After the sample stage, a detector is set up to record the reflected the signal. A thorough description of the measurement procedures (detector scan, sample height scan, geometric alignment, data processing) have been discussed in detail elsewhere and will not be addressed here [17]. Here, we mainly sketch the procedure to measure the reflectivity as a function of potential. First, we need a flat liquid/liquid interface. This is accomplished by pinning the interface to the top of a Teflon strip, where the latter is pressed to the inner glass wall of the sample cell by a strip of stainless steel shim stock. The interface is flattened by adjusting the volume of DCE phase. After acceptable sample height scans are produced, the Q-space geometry is defined. The reflectometer is moved into a specific Q_z point at which a series of electrostatic potentials are measured within the polarization window measured electrochemically (see Chap. 3). During the experiment, the current is monitored, while the open circuit potential reflectivity data is used as a reference measurement to insure that the system is completely reversible to polarization. This same procedure is then repeated at a different Q_z. The procedure to apply the potential is as follows: starting from some applied potential $\Delta\phi_{cell}^1$ usually chosen to be the open circuit potential, the potential is ramped up to the desired $\Delta\phi_{cell}^2$ using a rate of a few mV/s. Once $\Delta\phi_{cell}^2$ is reached, the system's current is allowed to equilibrate, reaching a constant value typically less than 1 μA. Afterwards, the data collection begins. This procedure of measuring reflectivity ensures that the data is reproducible, as confirmed for each data set on a different but identically prepared sample.

4.2 Data Analysis

Due to the limited Q_z range measurable at the liquid/liquid interface, standard routine analyses based on fitting of the electron density (model-dependent, model-independent, Patterson function, etc ...) would not provide us with as much information as direct comparison to some theoretical method. In addition, since reflectivity is not sensitive to the individual ions only to the overall electron density, the previously mentioned analyses would not produce ion distributions from which we can properly investigate the physics of the electrical double layer. Hence, we choose to directly

4.2 Data Analysis

compare the predictions of electrical double layer models to the data by calculating ion distributions from the models and converting them to an electron density profile from which the reflectivity can be readily obtained using the Parratt method [18]. This methodology was used by Luo et al. in [12, 13].

Fitting to the data involves calculating the electron density $\rho(z)$ and R/R_F from the ion concentration profiles $c_i(z)$, using the method described in [12]. The first step in this method is to define an *intrinsic* electron density profile, where the solvents are assumed to be homogeneously distributed,

$$\rho_{\text{solvent}}^{int}(z) = \begin{cases} \rho_w, & z > 0, \\ \rho_o, & z < 0, \end{cases} \quad (4.1)$$

with the following electron densities for water: $\rho_w = 0.33\,e^-/\text{Å}^3$, and DCE: $\rho_o = 0.38\,e^-/\text{Å}^3$. The electron density of the double layer is accounted for by smearing the charge of an ion throughout its volume using a Gaussian function, for this purpose the ions were modeled as spheres of diameter 1.95 Å for Na^+, 3.66 Å for Cl^-, 11.0 Å for $BTPPA^+$, and 10.0 Å for $TPFB^-$, where the latter were calculated from the crystal structure of BTPPATPFB. At some potential $\Delta\phi$, a theoretical model (see main text) is used to predict a density profile $n_i(z)$ of each ion i present in the system, the overall intrinsic electron density is given by,

$$\rho^{int}(z) = \rho_{\text{solvent}}^{int}(z) + \sum_i \left(\bar{n}_i(z) N_i^e - \bar{n}_i(z) V_i \rho_{\text{solvent}}^{int}(z) \right) \quad (4.2)$$

where the sum is over the ion types, $\bar{n}_i(z)$ is the size-smeared density profile of ion i, N_i^e is the number of electrons of ion i. While the first term in the sum counts the contribution of the ion densities to the electron density profile, the last term subtracts the solvent electron density within the ionic volume V_i to eliminate over counting of the solvent's electron density. The intrinsic electron density $\rho^{int}(z)$ is not directly measurable by reflectivity. Due to interfacial thermal fluctuations, reflectivity measures fluctuations superposed on the intrinsic electron density profile given in (4.2). Hence, the relevant electron density profile averaged over the xy-plane is given by

$$\langle \rho(z) \rangle_{xy} = \left(\rho^{int} \star f \right)(z) \quad (4.3)$$

where $\star$ is the convolution operation, and f is the probability density function assigned to the interfacial fluctuations. A useful model to describe interfacial fluctuations is capillary wave theory, which describes the latter as density fluctuations that are driven by thermal fluctuations and opposed by surface tension and gravity. This model provides a useful connection between the interfacial tension and rms value of height fluctuations $\langle h_{xy} \rangle$ or interfacial width, σ_{cap}

$$\sigma_{cap}^2(\phi) \equiv \langle h_{xy}^2 \rangle = \frac{k_B T}{2\pi \gamma(\phi)} \ln \frac{Q_{xy}^{max}}{Q_{xy}^{min}} \tag{4.4}$$

where $\gamma(\phi)$ is the potential-dependent interfacial tension (see Chap. 3), Q^{max} represents the maximum wavevector above which a continuum description of fluctuations breaks down (taken to be $\pi/10\,\text{Å}^{-1}$) and Q^{min} is the in-plane experimental resolution ($\approx 10^{-4}\,\text{Å}^{-1}$). The probability density function of the capillary wave model is given by the normal distribution $\mathcal{N}(z_o, \sigma)$ whose variance is determined by (4.4). Therefore in (4.3), we use

$$f = \mathcal{N}(0, \sigma_{cap}(\phi)) \tag{4.5}$$

Due to the dependence of the interfacial width and the ion distributions on the electrostatic potential, the procedure to define $\rho(z)$ is done at each potential. The x-ray reflectivity is calculated by Parratt's method [19], where the electron density profile is divided up into very small segments along the interfacial normal. The reflection and transmission coefficients can then be exactly calculated in each segment (including the effects of absorption) and the reflectivity is then obtained over the entire domain of electron density variation [17]. Fits of R/R_F to predictions of GC model (Fig. 4.3) used only the interfacial roughness (fitted values are listed in Table 4.1) and a Q_z offset ($10^{-4}\,\text{Å}^{-1}$, a typical misalignment of the reflectometer) as fitting parameters. These fits agree with the data at small $\Delta\phi$ (-0.12 to 0.18 V), but at larger $\Delta\phi$ (0.28 and 0.33 V), R/R_F is greatly overestimated primarily because Gouy-Chapman theory predicts large TPFB$^-$ ion concentrations near the interface, which provides the dominant x-ray contrast.

As discussed in Chap. 2, one can include the contributions of interactions omitted in the Poisson-Boltzmann theory by approximating the energy of ion i that enters

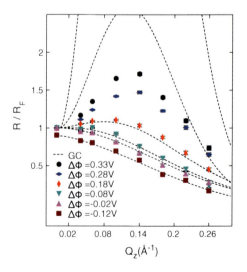

Fig. 4.3 Gouy-Chapman predictions of R/R_F compared to the data at various potentials

4.2 Data Analysis

Table 4.1 The potential dependent interfacial roughness of the 100 mM NaCl system: capillary wave theory and fits to the x-ray data

	Capillary wave theory (Å)	GC[a] (±0.20 Å)	PMF I (±0.20 Å)	PMF II (±0.20 Å)
$\Delta\phi = 0.33$ V	5.06	5.06	4.83	4.32
$\Delta\phi = 0.28$ V	4.83	4.83	4.73	4.34
$\Delta\phi = 0.18$ V	4.69	4.03	5.10	4.74
$\Delta\phi = 0.08$ V	4.51	4.22	5.17	5.08
$\Delta\phi = -0.02$ V	4.45	4.27	4.98	4.91
$\Delta\phi = -0.12$ V	4.45	3.84	4.78	4.83

[a]Roughness at $\Delta\phi = 0.33, 0.28$ V is not fitted for the GC model

the Boltzmann factor by $E_i(z) \approx e_i\phi(z) + W_i(z)$, where $W_i(z)$ is the potential of mean force (PMF) for each ion i. The resultant equation is

$$\frac{d^2\phi(z)}{dz^2} = -\frac{1}{\varepsilon_o \varepsilon} \sum_i e_i n_i^b \exp[-\beta(e_i\phi(z) + W_i(z))], \quad (4.6)$$

Since ion–ion correlations are negligible for monovalent aqueous ions, we have chosen to determine the PMF of Na$^+$ from a molecular dynamics (MD) simulation for a single ion (see Sect. 6.2). The PMF of Cl$^-$ was taken from an MD simulation in the literature [20]. Figure 4.4 illustrates the monotonic variation of $W_i(z)$ for Na$^+$ and Cl$^-$. Due to the computational difficulties of simulating $W_i(z)$ for large molecular ions such as BTPPA$^+$ and TPFB$^-$ (see Sect. 6.3), we used a phenomenological PMF previously introduced in [12, 13],

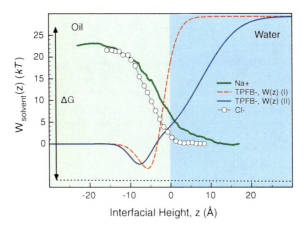

Fig. 4.4 Potentials of mean force for BTPPA$^+$ (*black*) and TPFB$^-$. $W^I_{\text{TPFB}-}$:*red*, $W^{II}_{\text{TPFB}-}$:*blue*, determined by fitting the reflectivity data in Fig. 4.5. PMFs for Na$^+$ (*green dots*) and Cl$^-$ (*circles*) were calculated by MD simulations (see text)

$$W_i(z) = (W_i(0) - W_i^p) \frac{\text{erfc}\left(|z| - \delta_i^p/L_i^p\right)}{\text{erfc}\left(-\delta_i^p/L_i^p\right)} + W_i^p, \qquad (4.7)$$

where $p(=\text{w, o})$ refers to either the water phase ($z \geq 0$) or the oil phase (DCE, $z \leq 0$), $W_i^o - W_i^w$ is the Gibbs energy of transfer of ion i from water to oil, δ_i^p is an offset to ensure continuity of $W_i(z)$ at $z = 0$, and L_i^p characterizes the decay of $W_i(z = 0)$ to its bulk values W_i^w and W_i^o. We used this monotonic PMF for BTPPA$^+$, but had to modify it for TPFB$^-$, as described below. Since $W_i^o - W_i^w$ for BTPPA$^+$ is known (Table 3.1), the PMF of BTPPA$^+$ is characterized by 3 parameters: $L^w_{\text{BTPPA}^+}$, $L^o_{\text{BTPPA}^+}$, $W_{\text{BTPPA}^+}(0)$ listed in Table 4.2. We determine the latter by fitting to R/R_F data at $\Delta\phi = -0.12$ V where it is expected that the BTPPA$^+$ interfacial concentration is enhanced. The large error bars on the PMF of BTPPA$^+$ are due to the small magnitude of the most negative $\Delta\phi^{w-o}$ that we studied. Moreover, the x-ray contrast of BTPPA$^+$ is not much different than DCE, $\rho_{BTPPA^+} = 0.39$ e$^-/$Å^3, further limiting precise determination of the parameters. The PMF fitting routine is described in detail in [21]. Briefly, parameters to define $W_i(z)$ are chosen from some parameter intervals, then the PB-PMF equation (4.6) is numerically solved, the reflectivity is computed from the predicted ion distributions as described earlier, and the reflectivity curve is compared to the data by calculating the χ^2 value, which measures the quality of a fit given uncertainties in the data. This procedure is then repeated to scan the entire parameter space to locate $W_i(z)$ with the minimum χ^2.

The x-ray reflectivity at the two highest positive potentials cannot be fit if (4.7) is used to model the PMF for TPFB$^-$. The simplest model that will produce the peaks in Fig. 4.5 is a single layer of TPFB$^-$ ions at the interface (note that a layer of Na$^+$, whose concentration is also enhanced at the interface, cannot provide the x-ray contrast required to fit the data). The TPFB$^-$ layer is modeled by an attractive well in the PMF. $W_{TPFB^-}(z)$ is given by (4.7) plus a Gaussian function $D \exp[-(z - z_0)^2/2\sigma_{PMF}^2]$ for $z < 0$ along with a constant offset at $z = 0$ to maintain continuity (see Fig. 4.4).

The six parameters of $W_{TPFB^-}(z)$ [z_0, D, σ_{PMF}, $L^w_{TPFB^-}$, $L^o_{TPFB^-}$, $W_{TPFB^-}(0)$] along with the Q_z offset and the interfacial roughness (4.3Å $< \sigma <$ 5.1Å) are determined by fitting R/R_F measured at $\Delta\phi^{w-o} = 0.28$ and 0.33 V, where the concentration of TPFB$^-$ is enhanced at the interface, shown in Tables 5.1 and 4.2. This fitting is performed under the constraint that the resultant $W_{ion}(z)$ produces R/R_F in agreement with the data over the *entire* range of potentials. The assumption of $W(z)$ being independent of $\phi(z)$ will be discussed at the end of the section. In addition, fitted PMFs were rejected if the fit value of the roughness σ was unphysically

Table 4.2 PMF parameters of BTPPA$^+$ and TPFB$^-$ obtained by fitting to the reflectivity data

	$W(0)$ (k_BT)	L^o (Å)	L^w (Å)	z_0 (Å)	σ_{PMF} (Å)	D (k_BT)
$W_{\text{BTPPA}^+}(z)$	14 +12/−6	20 +11/−6	13 ± 2	–	–	–
$W^I_{\text{TPFB}^-}(z)$	−5 ± 0.5	3 ± 0.1	9 ± 4	−3.5 ± 0.2	3.4 ± 0.2	−9 ± 0.25
$W^{II}_{\text{TPFB}^-}(z)$	−25 ± 0.5	11 ± 0.5	10 ± 2.7	−7.5 ± 0.3	2.6 ± 0.3	−5.25 ± 0.2

4.2 Data Analysis

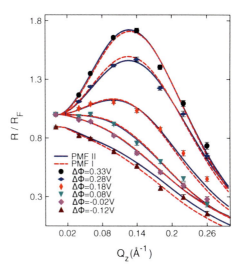

Fig. 4.5 PB-PMF predictions of R/R_F compared to the data at various potentials

small. In those cases an interfacial bending modulus [22] on the order of $100\,k_BT$ would have been required to reconcile the discrepancy of σ with its value predicted by capillary wave theory [23]. In the case of the $TPFB^-$ PMF, two local minima in χ^2-space (denoted $W^{I}_{TPFB^-}(z)$ and $W^{II}_{TPFB^-}(z)$) were found to satisfy these conditions. Potential profiles that are intermediate between $W^{I}_{TPFB^-}(z)$ and $W^{II(z)}_{TPFB^-}$ do not satisfy these conditions. Most of these fits had values of σ within one standard deviation of capillary wave theory predictions using the measured potential-dependent interfacial tension [12, 13]. Fits to $W^{II}_{TPFB^-}(z)$ at $\Delta\phi^{w-o} = 0.28$ and 0.33 V had values of σ within two standard deviations of capillary wave theory (Table 5.1).

The PB-PMF model with the $W_i(z)$ shown in Fig. 4.4 produces R/R_F in good agreement with the data over the entire range of measured potentials (Fig. 4.5). The attractive wells for $W^{I,II}_{TPFB^-}(z)$ have comparable depths (6 k_BT for $W^{I}_{TPFB^-}(z)$ and 5 k_BT for $W^{II}_{TPFB^-}(z)$), FWHM, and centers (Table 3.1). The ion concentration profiles $c_i(z)$, are calculated from (4.6) using $W_i(z)$. Figure 4.6 shows that the $c_i(z)$ at the highest potential, $\Delta\phi^{w-o} = 0.33$ V, take the form of two back-to-back double layers with a sharply defined layer of $TPFB^-$. The bulk concentration in both phases is constrained causing the ion density profiles to be discontinuous at the interface. The distribution of Na^+ near the interface is in stark contrast to that predicted from the Gouy-Chapman model, but agrees with the picture of a strongly hydrated aqueous ion at a water/oil interface, i.e Na^+ density profile increases as a consequence of the electrostatic interaction, yet prefers to remain hydrated, causing a depletion in its density in the immediate vicinity of the oil phase (Fig. 4.7). The different $c_i(z)$ calculated from $W^{I}_{TPFB^-}(z)$ or $W^{II}_{TPFB^-}(z)$ differ mainly in the broadness of the profile, which in the case of $W^{I}_{TPFB^-}(z)$ returns to its bulk value at $z = 0$, while $W^{II}_{TPFB^-}(z)$ allows $TPFB^-$ to penetrate slightly more into the water phase. The electron density

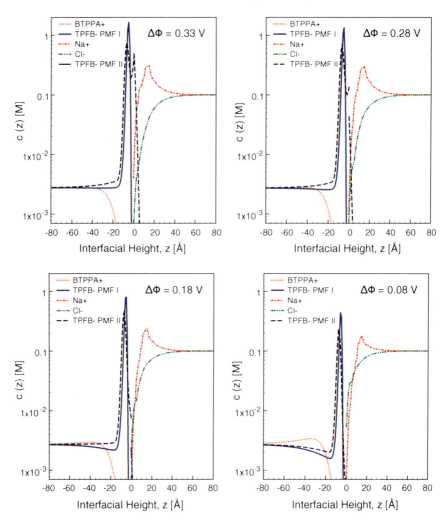

Fig. 4.6 PB-PMF ion concentration profiles as a function of positive potentials for the 100 mM NaCl (w)/5 mM BTPPATPFB (o) system

profiles $\rho(z)$ calculated from the different $c_i(z)$ are almost identical (Fig. 4.8), which demonstrates why our data cannot discriminate between $W^I_{TPFB^-}(z)$ and $W^{II}_{TPFB^-}(z)$.

The maximum density of TPFB$^-$ near the interface occurs at $\Delta\phi^{w-o} = 0.33$ V and is 1 nm^2 per TPFB$^-$ ion when $W^I_{TPFB^-}(z)$ is used or 1.5 nm^2 per TPFB$^-$ ion when $W^{II}_{TPFB^-}(z)$ is used. Both values represent a high-density layer for an ion of 1 nm diameter. Although dense ionic layers have been observed in the interfacial adsorption of charged amphiphiles [24], the absence of a dense TPFB$^-$ layer at $\Delta\phi^{w-o} \approx 0$ indicates that TPFB$^-$ is, at most, weakly amphiphilic, further supported by measurement of the Gibbs adsorption isotherm [21].

4.2 Data Analysis

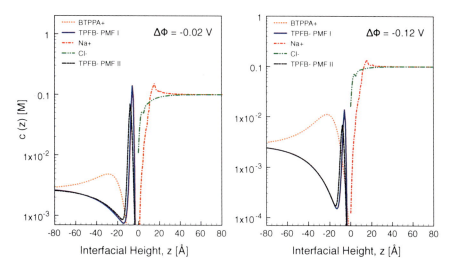

Fig. 4.7 PB-PMF ion concentration profiles as a function of negative potentials for the 100 mM NaCl (w)/5 mM BTPPATPFB (o) system

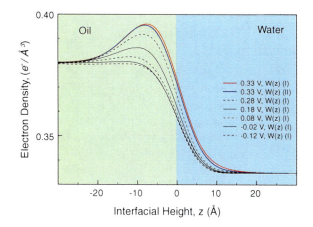

Fig. 4.8 Electron density profiles for various potentials calculated from PB-PMF. *Top* to *bottom*: $\Delta\phi = 0.33$ V, $W^{I}_{\text{TPFB}^-}$:*red*, $W^{II}_{\text{TPFB}^-}$:*blue*, 0.28 V (*dashed*), 0.18 V (*solid*), 0.08 V (*dashed*), -0.02 V (*solid*), and -0.12 V (*dashed*)

The MD simulations of the potentials of mean force that we used for Na^+ and Cl^- do not account for ion–ion correlations, but they do include ion–solvent and solvent–solvent correlations. Such correlations also account for the monotonic form of $W_{\text{BTPPA}^+}(z)$. However, as a result of modeling the x-ray reflectivity, the phenomenological $W_{\text{TPFB}^-}(z)$ in Fig. 4.4 must implicitly account for ion–ion correlations if they are important for the observed condensation. A deficiency of this phenomenological PB-PMF model is the lack of a physical mechanism that underlies this ionic layering. Hence, theoretical models will be studied in the next chapters to investigate the possible role of steric effects, which can stabilize large ion adsorption [25], and that of electrostatic ion correlations [26] in the observed ionic layering. The assumption of a $W_i(z)$ independent of the electrostatic potential is reasonable if solvent correlations dominate the PMF, within the limited polarization window. If ion–ion correlations are involved, $W_i(z)$ should depend on $\phi(z)$, although this is not the only instance where the latter can occur. In fact, such a result has been recently observed in the 10 mM LiCl data set, where we were able to apply a much larger electrostatic potential. A similar analysis as the one presented here indicated that a unique $W_{\text{TPFB}^-}(z)$ can fit the 10 mM LiCl data set up to $\Delta\phi = 0.4$ V, incidentally this is within the statistical uncertainties of $W_{\text{TPFB}^-}^{II}(z)$. However, when $\Delta\phi > 0.4$ V a statistically different PMF is needed to describe the data. Further details are presented in [21]. For the system studied in this chapter, no evidence of a PMF dependence on the potential is found within the polarization window. In Chap. 7 conclusive evidence is presented that the ion condensation presented here is due to electrostatic ion correlations, indicating that the assumption of a potential-independent $W(z)$ is only a first approximation to the interactions of ions in the system.

4.3 Concluding Remarks

The phenomena of monovalent ion condensation probed using the structural measurements of x-ray reflectivity is an interesting and novel result in the study of ion distributions. In the next three chapters, using theoretical models and Molecular Dynamics simulations we will utilize this observation as a stringent test of electrical double layer models and as a guide to understand the rich and complex interactions of ions in solution. Moreover, the PB-PMF model presented is shown to give a quantitative description of this ionic condensation when fitted against the XR data. The role that the potential of mean force $W_i(z)$ plays, is that of an excess chemical potential over the ideal gas approximation of the PB theory. Based on the results presented here and the previous results of [12, 13], the PB-PMF approach seems to provide a suitable approach to phenomenologically extend the Poisson-Boltzmann theory beyond the mean-field approach.

References

1. Als-Nielsen, J., Pershan, P.: Synchrotron X-ray diffraction study of liquid surfaces. Nucl. Inst. Meth. **208**, 545 (1983)
2. Braslau, A., Deutsch, M., Pershan, P., Weiss, A., Als-Nielsen, J., Bohr, J.: Surface roughness of water measured by X-ray reflectivity. Phys. Rev. Lett. **54**, 114 (1985)
3. Braslau, A., Pershan, P.S., Swislow, G., Ocko, B.M., Als-Nielsen, J.: Capillary waves on the surface of simple liquids measured by X-ray reflectivity. Phys. Rev. A **38**(5), 2457–2470 (Sep

References

1988)
4. Schwartz, D.K., Schlossman, M.L., Kawamoto, E.H., Kellogg, G.J., Pershan, P.S., Ocko, B.M.: Thermal diffuse X-ray-scattering studies of the water-vapor interface. Phys. Rev. A **41**(10), 5687–5690 (1990)
5. Sinha, S.K., Sirota, E.B., Garoff, S., Stanley, H.B.: X-ray and neutron scattering from rough surfaces. Phys. Rev. B **38**, 2297 (1988)
6. Als-Nielsen, J., McMorrow, D.: Elements of Modern X-ray Physics. Wiley, Hoboken (2001)
7. Daillant, J., Gibaud, A. (eds.): X-ray and Neutron Reflectivity: Principles and Applications, 2nd edn. Springer, Berlin (2010)
8. Mitrinovic, D.M., Tikhonov, A.M., Li, M., Huang, Z., Schlossman, M.L.: Noncapillary-wave structure at the water-alkane interface. Phys. Rev. Lett. **85**, 582 (2000)
9. Buff, F.P., Lovett, R.A., Stillinger, F.H.: Interfacial density profile for fluids in the critical region. Phys. Rev. Lett. **15**, 621 (1965)
10. Schlossman, M.L., Li, M., Mitrinovic, D.M., Tikhonov, A.M.: X-ray surface scattering studies of molecular ordering at liquid-liquid interfaces. In: Stock, S., Perry, D., Mini, S. (eds.) Applications of Synchrotron Radiation Techniques to Materials Science V, p. 165. Materials Research Society, Warrendale (2000)
11. Schlossman, M.L., Mitrinovic, D.M., Zhang, Z., Li, M., Huang, Z.: X-ray scattering from single liquid-liquid interfaces. Synchrotron Radiat. News **12**, 53–58 (1999)
12. Luo, G., Malkova, S., Yoon, J., Schultz, D.G., Lin, B., Meron, M., Benjamin, I., Vanysek, P., Schlossman, M.L.: Ion distributions near a liquid-liquid interface. Science **311**, 216–218 (2006)
13. Luo, G., Malkova, S., Yoon, J., Schultz, D.G., Lin, B., Meron, M., Benjamin, I., Vanysek, P., Schlossman, M.L.: Ion distributions at the nitrobenzene-water interface electrified by a common ion. J. Electroanal. Chem. **593**, 142–158 (2006)
14. Laanait, N., Yoon, J., Hou, B., Vanysek, P., Meron, M., Lin, B., Luo, G., Benjamin, I., Schlossman, M.: Communications: monovalent ion condensations at the electrified liquid/liquid interface. J. Chem. Phys. **132**, 171101 (2010)
15. Fenter, P., Sturchio, N.C.: Mineral-water interfacial structures revealed by synchrotron X-ray scattering. Prog. Surf. Sci. **77**(5–8), 171–258 (2004)
16. Schlossman, M.L., Synal, D., Guan, Y., Meron, M., Shea-McCarthy, G., Huang, Z., Acero, A., Williams, S.M., Rice, S.A., Viccaro, P.J.: A synchrotron X-ray liquid surface spectrometer. Rev. Sci. Instrum. **68**, 4372–4384 (1997)
17. Pershan, P. Schlossman, M.: Liquid Surfaces and Interfaces: Synchrotron X-ray Methods. Cambridge University Press, Cambridge (2012)
18. Paratt, L.: Surface studies of solids by total reflection of X-rays. Phys. Rev. **95**(2), 359–369 (1954)
19. Parratt, L.G.: Surface studies of solids by total reflection of X-rays. Phys. Rev. **95**, 359 (1954)
20. Wick, C.D., Dang, L.X.: Recent advances in understanding transfer ions across aqueous interfaces. Chem. Phys. Lett. **458**, 1–5 (2008)
21. Hou, B.: Ion distributions at electrified liquid-liquid interfaces: an agreement between X-ray reflectivity analysis and macroscopic measurements. Doctoral dissertation, University of Illinois at Chicago (2011)
22. Safran, S.A.: Statistical Thermodynamics of Surfaces, Interfaces, and Membranes. Addison-Wesley Publishing Co., Reading (1994)
23. Luo, G., Malkova, S., Pingali, S.V., Schultz, D.G., Lin, B., Meron, M., Benjamin, I., Vanysek, P., Schlossman, M.L.: Structure of the interface between two polar liquids: nitrobenzene and water. J. Phys. Chem. B **110**, 4527–4530 (2006)
24. Leveiller, F., Jacquemain, D., Lahav, M., Leiserowitz, L., Deutsch, M., Kjaer, K., Als-Nielsen, J.: Crystallinity of the double layer of Cadmium arachidate films at the water surface. Science **252**, 1532 (1991)
25. Borukhov, I., Andelman, D., Orland, H.: Steric effects in electrolytes: a modified Poisson-Boltzmann equation. Phys. Rev. Lett. **79**, 435 (1997)
26. Groot, R.D.: Ion condensation on solid particles: theory and simulations. J. Chem. Phys. **95**(12), 9191 (1991)

Chapter 5
Sterically Modified Poisson-Boltzmann Equation

Steric effects in charged soft-matter play important roles in numerous settings such as colloidal suspensions and greatly influence the behavior of macromolecular ions in solution. The Poisson-Boltzmann equation ignores these excluded-volume interactions, which prompted a number of researchers to address this deficiency at various levels of sophistication and theory, [1–6]. Recent work by Antypov et al. [7], reviews numerous approaches to address steric effects that are amenable to a functional free energy formulation, and compares their predictions to those of Monte Carlo simulations. The results of [7] seem to favor approaches based on nonlocal density functional theories. However, the correct methodology to account for these excluded-volume interactions is still under debate. Specifically, the interplay between excluded-volume correlations and electrostatic correlations remains an issue of contention that is currently addressed theoretically in a satisfactory manner only within ab initio approaches (those with an explicit treatment of the 2-body interatomic potentials), such as the anisotropic hypernetted chain [8]. Moreover, partially due to the close connection between size correlations and electrostatic correlations, experimental studies that probe ion distributions have only recently begun to address the theoretical questions concerning excluded-volume interactions of ions in solution.

In what follows we present a theoretical approach that aims to describe volume-excluded interactions in charged soft-matter, at the mean-field level, thereby ignoring ion correlations. This theory results in a sterically modified Poisson-Boltzmann (SPB) equation, and reduces to the Poisson-Boltzmann equation in the limit of zero size.

The SPB theory that we discuss, due to Borukhov et al., is based on a lattice Coulomb gas formalism [9] and utilizes a phenomenological free energy functional [9, 10]. In contrast to the PB result that the ion density is unbounded (see for example Fig. 5.2), the main prediction of the SPB theory is that near highly charged surfaces the counterions would exhibit a saturated ion density profile $n(\mathbf{r})$ at the close packing density of $1/a^3$, where a is the ion size, effectively stabilizing ionic condensation at the surface. This can be interpreted physically to be due to the loss of the solvent entropy, which counteracts the gain in electrostatic energy due to ion layering at the surface.

In fact, it will be seen that SPB theory is just the PB theory with the additional contribution of solvent entropy to the free energy functional, $\mathcal{F}_{PB}[n(\mathbf{r})]$ (see Chap. 2). In Sect. 5.1, we give a rigorous exposition of the field theory formalism used to derive $\mathcal{F}_{SPB}[n(\mathbf{r})]$ from the lattice Coulomb gas. In Sect. 5.2, a phenomenological free energy is postulated and shown to be equivalent to the lattice gas formalism. The SPB equation is then obtained through a variational procedure. We generalize Borukhov's SPB, in Sect. 5.3, to include excluded volume effects due to multiple ionic sizes. In Sect. 5.4, we discuss a numerical procedure to solve the SPB equation formulated as a boundary value problem for application to a liquid/liquid interface, and more generally models of ion distributions whose equations of motion are second order nonlinear ordinary differential equations. The predictions of both the simple SPB and the generalized SPB in a number of settings are also presented. In the last section, we compare x-ray reflectivity and thermodynamic data from the liquid/liquid interface to the results of SPB theory and show that the latter gives a qualitatively correct description of ion properties for this system when moderately charged.

5.1 Lattice Gas Approach

In this section, we follow the approach of Borukhov [9] in deriving the SPB equation. Consider a system composed of ions of charge $\pm ze$ and size a coupled to a bulk reservoir (a $z:z$ electrolyte) in a continuum dielectric medium of permitivity ε. The volume occupied by the system is divided into a three dimensional lattice, where each lattice cell has dimensions $a \times a \times a$, shown in Fig. 5.1. Restricting the occupation of each lattice site to a single ion produces a short-range repulsion. We introduce spin variables $\{s_i\}$ to denote the occupation of cell i, where $s_i = 1, -1, 0$ signifies that the lattice is occupied by a cation $(+)$, or an anion $(-)$, or unoccupied, respectively. With the ions only interacting through a Coulomb field $v_c = 1/\varepsilon|\mathbf{r}|$ (we use cgs units throughout), the Hamiltonian of the system written in terms of $\{s_i\}$ is:

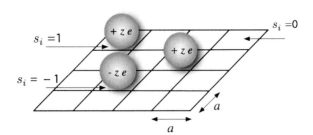

Fig. 5.1 Lattice gas modeling of $z{:}z$ electrolyte with lattice spacing a. The ions are represented as *spheres* with diameter a and charge $\pm ze$. The "spin" variables s_i indicate the occupancy of a lattice site

5.1 Lattice Gas Approach

$$H = \frac{1}{2}(ze)^2 \sum_{j,j'} s_j v_c(\mathbf{r}_j - \mathbf{r}_{j'}) s_{j'} - \sum_j \mu_j s_j^2 \qquad (5.1)$$

where $\mathbf{r}_i$ denotes the position of cell i, and μ_i is the chemical potential that couples the lattice site i to a bulk reservoir of ($\pm$) ions. The canonical partition function, is given by

$$Z = \mathrm{Tr}(e^{-\beta H}) \qquad (5.2)$$

$$= \sum_{s_i=1,-1,0} \exp\left(\frac{-\beta}{2}(ze)^2 \sum_{j,j'} s_j v_c(\mathbf{r}_j - \mathbf{r}_{j'}) s_{j'} + \beta \sum_j \mu_j s_j^2\right) \qquad (5.3)$$

To convert the discrete lattice cell occupation sums into spatial integrals, a field representation of the ion density is introduced through a charge density operator $\hat{\rho}$, defined by

$$\hat{\rho}_c(\mathbf{r}) = \sum_j z e s_j \delta(\mathbf{r} - \mathbf{r}_j) \qquad (5.4)$$

where δ is Dirac's delta function. The charge density operator, when acting on an ion number density $\rho(\mathbf{r}_i)$ associated with cell i, gives the charge density at $\mathbf{r}_i$, as follows

$$\int d\mathbf{r}_i \hat{\rho}_c(\mathbf{r}) \rho(\mathbf{r}_i) = \sum_j z e s_j \int d\mathbf{r}_i \delta(\mathbf{r} - \mathbf{r}_j) \rho(\mathbf{r}_i) \qquad (5.5)$$

$$= \sum_j z e s_j \delta_{ij} \rho(\mathbf{r}_i) \qquad (5.6)$$

$$= z e s_i \rho(\mathbf{r}_i) \qquad (5.7)$$

the definition of the Dirac delta function is used from the 1st to 2nd line, where δ_{ij} is the Kronecker delta. Global charge neutrality is trivially satisfied by taking the integral of $\hat{\rho}_c(\mathbf{r})$ over all space,

$$\int d\mathbf{r}_i \hat{\rho}_c(\mathbf{r}) = \sum_j z e s_j = 0 \qquad (5.8)$$

Using the definitions above, we can rewrite the sums over the spin variables in (5.3), in terms of the charge density operator,

$$Z = \sum_{s_i} \exp\left(-\frac{\beta}{2} \int d\mathbf{r} d\mathbf{r}' \hat{\rho}_c(\mathbf{r}) v_c(\mathbf{r} - \mathbf{r}') \hat{\rho}_c(\mathbf{r}') + \beta \sum_j \mu_j s_j^2\right) \qquad (5.9)$$

$$= \exp\left(-\frac{\beta}{2} \int d\mathbf{r} d\mathbf{r}' \hat{\rho}_c(\mathbf{r}) v_c(\mathbf{r} - \mathbf{r}') \hat{\rho}_c(\mathbf{r}')\right) \times \sum_{s_i} e^{\beta \sum_j \mu_j s_j^2} \qquad (5.10)$$

The double integral in (5.10) cannot be performed analytically, because the particles are coupled via the 2-body potential, v_c. To linearize the charge density fields, we need to introduce an auxiliary field, denoted by φ_c, then use a Hubbard-Stratonovich transformation [11] to map the system into a system of particles that interact only with this auxiliary field. For this theory, the Hubbard-Stratonovich transformation is equivalent to the following identity,

$$1 = \int \mathcal{D}\rho_c \mathcal{D}\varphi_c \exp\left(i\int d\mathbf{r} \rho_c(\mathbf{r})\varphi_c(\mathbf{r}) - ize \sum_j s_j \varphi_c(\mathbf{r})\right) \quad (5.11)$$

where we have set $\beta = 1$, to be restored when we derive the free energy. The above identity, is easily derived from the spectral definition of the Dirac delta function, $\delta(\mathbf{r}-\mathbf{r}') \propto \int d\mathbf{k} e^{i\mathbf{k}\cdot(\mathbf{r}-\mathbf{r}')}$, generalized to a functional (up to a constant which cancels out at the end of the derivation),

$$\delta[\rho_c(\mathbf{r}_i) - \hat{\rho}_c(\mathbf{r})] = \int d\varphi_c(\mathbf{r}_i) e^{i\varphi_c(\mathbf{r}_i)(\rho_c(\mathbf{r}_i) - \hat{\rho}_c(\mathbf{r}))} \quad (5.12)$$

Extending the above definition for all $\{\mathbf{r}_i\}$, we have

$$\delta[\rho_c(\mathbf{r}) - \hat{\rho}_c(\mathbf{r})] = \prod_i \delta[\rho_c(\mathbf{r}_i) - \hat{\rho}_c(\mathbf{r})] \quad (5.13)$$

$$= \prod_i \int d\varphi_c(\mathbf{r}_i) e^{i\varphi_c(\mathbf{r}_i)(\rho_c(\mathbf{r}_i) - \hat{\rho}_c(\mathbf{r}))} \quad (5.14)$$

$$= \int\int \cdots d\varphi_c(\mathbf{r}_1) d\varphi_c(\mathbf{r}_2) \cdots e^{i\varphi_c(\mathbf{r}_1)(\rho_c(\mathbf{r}_1) - \hat{\rho}_c(\mathbf{r}))} e^{i\varphi_c(\mathbf{r}_2)}$$
$$\times (\rho_c(\mathbf{r}_2) - \hat{\rho}_c(\mathbf{r})) \cdots \quad (5.15)$$

$$= \int \mathcal{D}\varphi_c \exp\left(i \sum_i \varphi_c(\mathbf{r}_i)\rho_c(\mathbf{r}_i) - ize \sum_{ij} s_j \delta(\mathbf{r} - \mathbf{r}_i)\varphi_c(\mathbf{r}_i)\right) \quad (5.16)$$

$$\delta[\rho_c(\mathbf{r}) - \hat{\rho}_c(\mathbf{r})] = \int \mathcal{D}\varphi_c \exp\left(i \int d\mathbf{r}\varphi_c(\mathbf{r})\rho_c(\mathbf{r}) - ize \sum_j s_j \varphi_c(\mathbf{r}_j)\right) \quad (5.17)$$

In the 2nd line, (5.12) was used, in the 3rd line, the product of the exponential terms is summed. In the 4th line, we apply the definition of $\hat{\rho}_c$ and define the functional integral, i.e, $\prod_i \int d\varphi_c(\mathbf{r}_i) \to \int \mathcal{D}\varphi_c$. Finally, in the last line the discrete sum that does not involve a lattice site occupation is converted to an integral. The final step in the derivation of (5.11) is taking the integral of both sides of the last expression (incidentally canceling the proportionality constant we omitted),

5.1 Lattice Gas Approach

$$1 \equiv \int D\rho_c \delta[\rho_c(\mathbf{r}) - \hat{\rho}_c(\mathbf{r})]$$

$$= \int D\rho_c D\varphi_c \exp\left(i \int d\mathbf{r} \varphi_c(\mathbf{r})\rho_c(\mathbf{r}) - ize \sum_j s_j \varphi_c(\mathbf{r}_j)\right) \quad (5.18)$$

Inserting the above identity into (5.10), gives

$$Z = \int D\varphi_c \underbrace{\int D\rho_c \exp\left(-\frac{1}{2}\int d\mathbf{r}d\mathbf{r}' \hat{\rho}_c(\mathbf{r})v_c(\mathbf{r}-\mathbf{r}')\hat{\rho}_c(\mathbf{r}') + i\int d\mathbf{r}\rho_c(\mathbf{r})\varphi_c(\mathbf{r})\right)}_{\mathscr{A}(\varphi_c)} \quad (5.19)$$

$$\times \underbrace{\sum_{s_i} e^{-ize\sum_j s_j \varphi_c(\mathbf{r}) + \sum_j \mu_j s_j^2}}_{\mathscr{B}()} \quad (5.20)$$

where we have factorized the partition function into two terms, $\mathscr{A}(\varphi_c)$ contains functional integration over the density fields, and $\mathscr{B}()$ contains discrete sums over lattice site occupations. The partition function is given by:

$$Z = \int D\varphi_c \mathscr{A}(\varphi_c) \mathscr{B}(\varphi_c) \quad (5.21)$$

$\mathscr{B}(\varphi_c)$ can be manipulated to yield,

$$\mathscr{B}(\varphi_c) = \exp\left(\ln\left\{\sum_{s_i=1,-1,0} e^{-ize\sum_j s_j \varphi_c(\mathbf{r}) + \sum_j \mu_j s_j^2}\right\}\right) \quad (5.22)$$

$$= \exp\left(\ln\left\{\sum_{s_i=1,-1,0} \prod_i e^{-izes_i \varphi_c(\mathbf{r}_i) + \mu_i s_i^2}\right\}\right) \quad (5.23)$$

$$= \exp\left(\sum_i \ln\left\{\sum_{s_i=1,-1,0} e^{-izes_i \varphi_c(\mathbf{r}_i) + \mu_i s_i^2}\right\}\right) \quad (5.24)$$

Now, we can convert the summation over lattice cells into spatial integrals, $\sum_i \to 1/a^3 \int d\mathbf{r}$. Putting this into the expression above, and performing the summation over the cell occupation, we obtain

$$\mathscr{B}(\varphi_c) = \exp\left(\frac{1}{a^3}\int d\mathbf{r} \ln[1 + e^{\mu_+ - ize\varphi_c(\mathbf{r})} + e^{\mu_- + ize\varphi_c(\mathbf{r})}]\right) \quad (5.25)$$

with obvious notation for the chemical potential of $\pm$ ions.

The $\mathscr{A}(\varphi_c)$ term requires more involved calculations than $\mathscr{B}()$, in order to evaluate the path integrals. However, there are standard field-theoretic methods that one can follow [12]. We start by writing $\rho_c(\mathbf{r}_i) \equiv \rho_i$, $\varphi_c(\mathbf{r}_i) \equiv \varphi_i$ and $v_c(\mathbf{r}_i - \mathbf{r}_j) \equiv v_{ij}$. The density field (and its conjugate) ρ_i (φ_i) can be thought of as the i component of an infinite-dimensional column vector ρ (φ). While, v_{ij} represents the ij element of the (infinite dimensional) matrix Coulomb kernel, v_c. Recall that $\mathscr{A}(\varphi_c)$ is given by,

$$\mathscr{A}(\varphi_c) = \int \mathcal{D}\rho_c \exp\left(-\frac{1}{2}\int \mathrm{d}\mathbf{r}\mathrm{d}\mathbf{r}'\hat{\rho}_c(\mathbf{r})v_c(\mathbf{r}-\mathbf{r}')\hat{\rho}_c(\mathbf{r}') + i\int \mathrm{d}\mathbf{r}\rho_c(\mathbf{r})\varphi_c(\mathbf{r})\right) \quad (5.26)$$

replacing the integrals with sums and using the above notation, we have

$$\mathscr{A}(\varphi_c) = \int \prod_l \int \mathrm{d}\rho_l \exp\left(-\frac{1}{2}\sum_{ij}\rho_i v_{ij}\rho_j - i\sum_i \rho_i \varphi_i\right) \quad (5.27)$$

Using matrix notation we could write the sums as follows:

$$\sum_{ij} \rho_i v_{ij} \rho_j = \rho^\mathrm{T} v_c \rho \quad \text{and} \quad \sum_i \rho_i \varphi_i = \rho^\mathrm{T} \varphi \quad (5.28)$$

where T stands for the transpose. The term quadratic in ρ is diagonalized by means of an orthogonal transformation $\rho' = O\rho$, where O is an orthogonal matrix ($O^\mathrm{T} = O^{-1}$). We now have,

$$\rho^\mathrm{T} v_c \rho = \rho'^\mathrm{T} O v_c O^\mathrm{T} \rho' = \rho'^\mathrm{T} \mathrm{diag}(v_c)\rho' \quad (5.29)$$

$$\rho^\mathrm{T} \varphi = \rho'^\mathrm{T} O \varphi \quad (5.30)$$

where diag( ) stands for diagonal, to represent that the Coulomb kernel is now in diagonal form with eigenvalues $\{\lambda_i\}$. Reverting to the index notation, $\mathscr{A}(\varphi_c)$ is given by

$$\mathscr{A}(\varphi_c) = \prod_l \int \det(O)\mathrm{d}\rho'_l \exp\left(-\frac{1}{2}\sum_i \lambda_i \rho_i'^2 - i\sum_{ik} \rho'_k O_{ki}\varphi_i\right) \quad (5.31)$$

where det( ) stands for determinant, the jacobian appears due to transformation of the integration measure $\mathrm{d}\rho$, but since O is an orthogonal matrix, then $\det(O) = 1$. Since the density fields are now decoupled, we can perform their functional integrals, when the exponential term is rewritten in the form of a Gaussian.

$$\mathscr{A}(\varphi_c) = \prod_l \int \mathrm{d}\rho_l \prod_i \exp\left(-\frac{1}{2}\lambda_i \rho_i^2 - i\sum_k \rho_k O_{ki}\varphi_i\right) \quad (5.32)$$

5.1 Lattice Gas Approach

$$= \prod_i \int d\rho_i \exp\left(-\frac{1}{2}\lambda_i \rho_i^2 - i\sum_k \rho_k O_{ki}\varphi_i\right) \tag{5.33}$$

$$= \prod_i \int d\rho_i \exp(-\frac{\lambda_i}{2}\{\rho_i + \frac{i}{\lambda_i}\sum_k O_{ki}\varphi_i\}^2 + \frac{\lambda_i}{2}(\frac{i}{\lambda_i}\sum_k O_{ki}\varphi_i)^2) \tag{5.34}$$

$$= \left(\prod_i \int d\rho_i \exp(-\frac{\lambda_i}{2}\{\rho_i + \frac{i}{\lambda_i}\sum_k O_{ki}\varphi_i\}^2)\right)$$

$$\times \underbrace{\exp\left(\sum_i \frac{\lambda_i}{2}(\frac{i}{\lambda_i}\sum_k O_{ki}\varphi_i)^2\right)}_{C(\varphi)} \tag{5.35}$$

In the 1st line, the sum is converted to product of exponentials, we also relabel ρ' by ρ. In the 2nd line, the product of exponentials and the product of integrals are expanded then factorized. In the 3rd line, we complete the square, and in the last line exponential terms that don't depend on ρ are taken out of the integral. To evaluate the remaining gaussian integrals, we shift $\rho_i + \frac{i}{\lambda_i}\sum_k O_{ki}\varphi_i \to \rho_i$, giving

$$\mathcal{A}(\varphi_c) = \prod_i \frac{\sqrt{2\pi}}{\sqrt{\lambda_i}} \times C(\varphi_c) \tag{5.36}$$

$$= \frac{\sqrt[K]{2\pi}}{\sqrt{\det(v_c)}} \times C(\varphi_c) \tag{5.37}$$

where we used the identity $\det(M) = \prod 1/\lambda_i$, when M is a diagonal matrix, and K is the number of ρ-fields. This overall multiplicative constant does not affect the physics, and is easily seen to cancel out in the expression of any physical quantity derived from the partition function. Now, we turn our attention to the term involving the fluctuating field, $C(\varphi_c)$. Using matrix notation,

$$C(\varphi_c) = \exp\left(-\frac{1}{2}\sum_i \lambda_i [\lambda_i^{-1}(O\varphi)^T][\lambda_i^{-1}(O\varphi)]\right) \tag{5.38}$$

$$= \exp\left(-\frac{1}{2}\sum_i \varphi^T O^T \lambda_i^{-1} O\varphi\right) \tag{5.39}$$

$$= \exp\left(-\frac{1}{2}\varphi^T O^T \text{diag}(v_c^{-1})O\varphi\right) \tag{5.40}$$

$$= \exp\left(-\frac{1}{2}\varphi^T v_c^{-1}\varphi\right) \tag{5.41}$$

$$= \exp\left(-\frac{1}{2}\int d\mathbf{r}d\mathbf{r}' \varphi_c(\mathbf{r}')v_c^{-1}(\mathbf{r}-\mathbf{r}')\varphi_c(\mathbf{r})\right) \tag{5.42}$$

In the 3rd line, we use matrix notation to convert the sum of the reciprocals of the eigenvalues to the inverse of the diagonalized Coulomb matrix. In the 4th line, we use the orthogonal matrix O and its inverse, to transform diag(v_c^{-1}) back to its non-diagonal form v_c^{-1}. In the last line, we use the continuous representation of the operators. The inverse of the Coulomb operator, $v_c^{-1}(\mathbf{r}-\mathbf{r}')$, is obtained from Poisson equation of a point charge located at $\mathbf{r}'$, $\nabla^2 \phi = -4\pi z e \delta(\mathbf{r}-\mathbf{r}')/\varepsilon$. The inverse kernel is given by, $v_c^{-1}(\mathbf{r}-\mathbf{r}') = -\varepsilon \nabla^2 \delta(\mathbf{r}-\mathbf{r}')/(4\pi(ze)^2)$. Plugging this result in $C(\varphi_c)$, (note that we omit the factor $(ze)^2$, since we defined $v_c = 1/\varepsilon|\mathbf{r}|$ with the ze absorbed in the definition of the density operator)

$$C(\varphi_c) = \exp\left(-\frac{1}{2}\int d\mathbf{r}d\mathbf{r}'\varphi_c(\mathbf{r})(\frac{-\varepsilon}{4\pi}\delta(\mathbf{r}-\mathbf{r}')\nabla^2)\varphi_c(\mathbf{r}')\right) \quad (5.43)$$

$$= \exp\left(\frac{\varepsilon}{8\pi}\int d\mathbf{r}\varphi_c(\mathbf{r})\nabla^2\varphi_c(\mathbf{r})\right) \quad (5.44)$$

$$= \exp\left(-\frac{\varepsilon}{8\pi}\int d\mathbf{r}|\nabla\varphi_c(\mathbf{r})|^2 - \varphi_c\nabla\varphi_c|_\infty\right) \quad (5.45)$$

In the 2nd line, we used the definition of the Dirac delta function. In the 3rd line, we integrated by parts. The boundary term is zero, since the field vanishes at infinity. Putting this expression of $C(\varphi_c)$ into (5.37), we have for $\mathscr{A}(\varphi_c)$

$$\mathscr{A}(\varphi_c) = \frac{\sqrt[N]{2\pi}}{\sqrt{\det(v_c)}}\exp\left(-\frac{\varepsilon}{8\pi}\int d\mathbf{r}|\nabla\varphi_c(\mathbf{r})|^2\right) \quad (5.46)$$

Finally, (5.46) and (5.25) are plugged in (5.21), and restoring β gives the partition function,

$$Z = \int \frac{\mathcal{D}\varphi_c}{\sqrt{\det(v_c)/(2\pi)^N}}$$
$$\exp\left(\int d\mathbf{r}\{-\frac{\beta\varepsilon}{8\pi}|\nabla\varphi_c(\mathbf{r})|^2 + \frac{1}{a^3}\ln[1 + e^{\beta\mu_+ - ize\beta\varphi_c(\mathbf{r})} + e^{\beta\mu_- + ize\beta\varphi_c(\mathbf{r})}]\}\right)$$
(5.47)

The theory of particles on a lattice interacting through a Coulomb potential has been converted to a field theory of the fluctuating field φ_c. Given our expression for the partition function, we can derive physical properties of the system, such as the number density, etc .... However, a close inspection of (5.47), reveals that the chemical potentials $\mu_\pm$ are undefined. We proceed to derive an expression for them in terms of the total number of particles in the system N. The final expression for the bulk chemical potential also provides a way to check the self-consistency of the statistical field theory defined by (5.47). The average number of ($\pm$)-ions is given by,

5.1 Lattice Gas Approach

$$N_\pm = \frac{1}{Z} \frac{\partial Z}{\partial (\beta \mu_\pm)} \tag{5.48}$$

$$= \frac{1}{Z} \int \frac{\mathcal{D}\varphi_c}{\sqrt{\det(v_c)/(2\pi)^N}}$$

$$\exp\left(\int d\mathbf{r}\{-\frac{\beta\varepsilon}{8\pi}|\nabla\varphi_c(\mathbf{r})|^2 + \frac{1}{a^3}\ln[1 + e^{\beta\mu_+ - ize\beta\varphi_c(\mathbf{r})} + e^{\beta\mu_- + ize\beta\varphi_c(\mathbf{r})}]\}\right) \tag{5.49}$$

$$\times \frac{1}{a^3}\int d\mathbf{r} \frac{e^{\beta\mu_\pm \mp ize\beta\varphi_c(\mathbf{r})}}{1 + e^{\beta\mu_+ - ize\beta\varphi_c(\mathbf{r})} + e^{\beta\mu_- + ize\beta\varphi_c(\mathbf{r})}} \tag{5.50}$$

$$N_\pm = \left\langle \frac{1}{a^3}\int d\mathbf{r} \frac{e^{\beta\mu_\pm \mp ize\beta\varphi_c(\mathbf{r})}}{1 + e^{\beta\mu_+ - ize\beta\varphi_c(\mathbf{r})} + e^{\beta\mu_- + ize\beta\varphi_c(\mathbf{r})}} \right\rangle \tag{5.51}$$

where $\langle \ldots \rangle$ stands for the canonical average. In the thermodynamic limit ($N \to \infty$, N/V finite), where N is the total number of particles in the system and V is the system's volume, the ensemble average is found by letting $\int d\mathbf{r} \to V$ and evaluating the integrand for a zero fluctuating field,

$$N_\pm = \frac{V}{a^3} \frac{e^{\beta\mu_\pm \mp ize\beta\varphi_c(\mathbf{r})}}{1 + e^{\beta\mu_+ - ize\beta\varphi_c(\mathbf{r})} + e^{\beta\mu_- + ize\beta\varphi_c(\mathbf{r})}}\bigg|_{\varphi_c=0} \tag{5.52}$$

$$N_\pm = \frac{V}{a^3} \frac{e^{\beta\mu_\pm}}{1 + e^{\beta\mu_+} + e^{\beta\mu_-}} \tag{5.53}$$

Due to global charge neutrality, we must have $N_+ = N_- = N/2$. Defining the volume fraction, $\eta_o = Na^3/V = 2n^b a^3$, where we have introduced the bulk density of the ions, $n^b = N_\pm/V$, the chemical potential as a function of η_o, is found by inverting (5.53),

$$\mu_+ = \mu_- = \frac{1}{\beta}\ln\left(\frac{1}{2}\frac{\eta_o}{1-\eta_o}\right) \tag{5.54}$$

As expected the bulk chemical potential is consistent with the physics of the lattice gas, this is clearly seen by expanding the logarithmic function in (5.54), and using the definition of η_o,

$$\mu_\pm = \frac{1}{\beta}(\ln(n^b a^3) - \ln(1 - \sum_{i=+,-} n_i^b a^3)) \tag{5.55}$$

The first term on the right hand side of the equation is just the chemical potential of an ideal gas, μ_{id}. However due to restricting the lattice site occupation, which generates repulsions, $\mu_\pm$ needs to be higher than μ_{id}, this is taken care of by the second term on the right hand side. Moreover, in the limit of infinite dilution ($n_b a^3 \ll 1$), we need to have $\mu_\pm = \mu_{id}$, as is the case in (5.54).

The functional integral in the partition function needs to evaluated. This is done using the stationary-phase approximation [13], where the integrand in (5.47) is evaluated at its saddle-point, $\phi(\mathbf{r}) = i\varphi_c(\mathbf{r})$. This is equivalent to the mean-field approximation, where the fluctuation field φ_c is replaced by the macroscopic average of the electrostatic potential, ϕ. We obtain,

$$Z|_{\phi(\mathbf{r})=i\varphi_c(\mathbf{r})}$$
$$= Z_{v_c} \exp\left(\int d\mathbf{r}\{\frac{\beta\varepsilon}{8\pi}|\nabla\phi(\mathbf{r})|^2 + \frac{1}{a^3}\ln[1 + e^{\beta\mu_+ - ze\beta\phi(\mathbf{r})} + e^{\beta\mu_- + ze\beta\phi(\mathbf{r})}]\}\right) \quad (5.56)$$

$$= Z_{v_c} \exp\left(\int d\mathbf{r}\{\frac{\beta\varepsilon}{8\pi}|\nabla\phi(\mathbf{r})|^2 + \frac{1}{a^3}\ln[1 + \frac{\eta_o}{1-\eta_o}\cosh(ze\beta\phi(\mathbf{r}))]\}\right) \quad (5.57)$$

where $Z_{v_c} \sim 1/\sqrt{\det(v_c)}$ is a constant (corresponding to the self-energy), and will be subtracted off when defining the free energy. Also, in the 2nd line we used (5.54). The free energy can now be derived from the partition function,

$$\mathcal{F}[\phi(\mathbf{r})] = -\frac{1}{\beta}(\ln Z - \ln Z_{v_c}) \quad (5.58)$$

$$= -\frac{\varepsilon}{8\pi}\int d\mathbf{r}|\nabla\phi(\mathbf{r})|^2 - \frac{1}{\beta a^3}\int d\mathbf{r}\ln[1 + \frac{\eta_o}{1-\eta_o}\cosh(ze\beta\phi(\mathbf{r}))] \quad (5.59)$$

We can now derive the equation of motion for the electrostatic potential, from the variation of $\mathcal{F}$ with respect to ϕ,

$$\frac{\delta\mathcal{F}}{\delta\phi(\mathbf{r})} = -\frac{\varepsilon}{8\pi}\int d\mathbf{r}'\frac{\delta|\nabla\phi(\mathbf{r}')|^2}{\delta\phi(\mathbf{r})} - \frac{ze}{a^3}\frac{\eta_o}{1-\eta_o}$$
$$\int d\mathbf{r}'\frac{\delta\phi(\mathbf{r}')}{\delta\phi(\mathbf{r})}\frac{\sinh(ze\beta\phi(\mathbf{r}'))}{1 + \frac{\eta_o}{1-\eta_o}\cosh(ze\beta\phi(\mathbf{r}'))} \quad (5.60)$$

Using the identity $\frac{\delta\phi(\mathbf{r}')}{\delta\phi(\mathbf{r})} = \delta(\mathbf{r}' - \mathbf{r})$ and the fact that $\frac{\delta|\nabla\phi(\mathbf{r})|^2}{\delta\phi(\mathbf{r})} = -2\delta(\mathbf{r}' - \mathbf{r})$ $\nabla^2\phi(\mathbf{r}')$ (A.5), derived in the appendix. We find,

$$\frac{\delta\mathcal{F}}{\delta\phi(\mathbf{r})} = \frac{\varepsilon}{4\pi}\nabla^2\phi(\mathbf{r}) - \frac{ze}{a^3}\frac{\eta_o}{1-\eta_o}\frac{\sinh(ze\beta\phi(\mathbf{r}))}{1 + \frac{\eta_o}{1-\eta_o}\cosh(ze\beta\phi(\mathbf{r}))} \quad (5.61)$$

Then, $\frac{\delta\mathcal{F}}{\delta\phi(\mathbf{r})} = 0$, implies that,

$$\nabla^2\phi(\mathbf{r}) = \frac{8\pi zen_b}{\varepsilon}\frac{\sinh(ze\beta\phi(\mathbf{r}))}{1 - \eta_o + \eta_o\cosh(ze\beta\phi(\mathbf{r}))} \quad (5.62)$$

5.1 Lattice Gas Approach

We call (5.62) the sterically modified Poisson-Boltzmann equation (SPB). The SPB model is a differential equation that is highly nonlinear in ϕ, therefore recourse to numerical methods is necessary to solve for the electrostatic potential distribution and the ion distributions (treated in Sect. 5.4). However, the main physical predictions of (5.62) can be extracted by taking a few limits. For instance, in the limit of zero size $\eta_o \to 0$, SPB reduces to the PB equation of a $z:z$ electrolyte. The SPB density profile is given by

$$n_\pm(\mathbf{r}) = n^b \frac{e^{\mp z e \beta \phi(\mathbf{r})}}{1 - \eta_o + \eta_o \cosh(z e \beta \phi(\mathbf{r}))} \quad (5.63)$$

implying that $n_\pm$ is bounded from above by $1/a^3$ for all values of ϕ. Moreover, in the limit of large electrostatic potentials, $ze\beta\phi \gg 1$, the counterion density behaves as $\frac{1}{a^3} \frac{1}{1 + \frac{2-\eta_o}{\eta_o} e^{-ze\beta\phi}}$, indicating that the density profile saturates as a function of $\phi(\mathbf{r})$, and does not reach the PB contact value (further discussion of this result is postponed to Sect. 5.6). Additionally, in the case of moderate (small) electrostatic potentials the SPB should predict ion density profile closely resembling those of the PB, for in that regime the density contact value is small and the steric effects are expected to be small (negligible). A comparison between the predictions of the SPB and PB equations for the counterions distribution are shown in Fig. 5.2, illustrating the various limits just discussed.

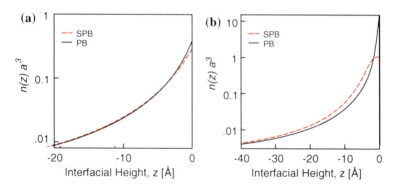

Fig. 5.2 Comparison of counterion density profiles predicted by the SPB and PB models. The numerical solutions of the models are obtained for a $1:1$ electrolyte, in a medium of dielectric constant $\varepsilon = 10$, $T = 300$ K. An ionic size of 11 Å was used for the SPB theory. The bulk ionic density is kept fixed at $n^b a^3 = 2 \times 10^{-3}$. **a** Electrostatic potential difference of $\Delta\phi = 5k_B T/e$ is applied, as discussed in the text the SPB and PB results are in close agreement. **b** Large electrostatic potential is applied, $\Delta\phi = 12k_B T/e$ showing the density saturation predicted by the SPB theory

5.2 Density Functional Approach

A simpler route to the SPB equation than the field-theory of the lattice Coulomb gas is a density functional theory (DFT). Following [10], the free energy functional of a $z:z$ electrolyte in a medium with dielectric constant ε is

$$\mathcal{F}[\phi(\mathbf{r}), n_\pm(\mathbf{r})] = U - TS \tag{5.64}$$

where U is the internal energy, T is the temperature and S is the entropy. The internal energy of the SPB model is

$$U_{SPB} = U_{PB}(\mu_{id} \to \mu_\pm) \tag{5.65}$$

where the expression of the PB internal energy is used after the replacement of the ideal gas chemical potential by its lattice gas counterpart (Eq. 5.54). The entropy functional of the lattice gas is,

$$S_{SPB} = \frac{-k_B}{a^3} \int d\mathbf{r}[n_+ a^3 \ln(n_+ a^3) + n_- a^3 \ln(n_- a^3) \\ + (1 - n_+ a^3 - n_- a^3)\ln(1 - n_+ a^3 - n_- a^3)] \tag{5.66}$$

The first and second term are the entropies of the positive and negative ions, respectively. The last term is due to restricting the occupation of a lattice site. To shed more light on (5.66), we take its limit far from close packing $n_\pm a^3 \ll 1$, by expanding the $\ln(1 - n_+ a^3 - n_- a^3)$ term to first order in $n_\pm$,

$$S_{SPB} \approx \frac{-k_B}{a^3} \int d\mathbf{r}[n_+ a^3 \ln(n_+ a^3) + n_- a^3 \ln(n_- a^3) \\ + (1 - n_+ a^3 - n_- a^3)(-n_+ a^3 - n_- a^3) + \mathcal{O}(n_\pm^2)] \tag{5.67}$$

$$= -k_B \int d\mathbf{r}[n_+ \ln(n_+ a^3) + n_- \ln(n_- a^3) - n_+ - n_- + \mathcal{O}(n_\pm^2)] \tag{5.68}$$

$$= -k_B \int d\mathbf{r}[n_+ (\ln(n_+ \Lambda^3) - 1) + n_- (\ln(n_- \Lambda^3) - 1)] \tag{5.69}$$

$$= S_{PB} \tag{5.70}$$

In the line before last, we replace a by the thermal wavelength Λ, given that in the infinite dilution limit the latter sets the relevant length scale. As we expected, the lattice gas entropy has the ideal gas entropy as a limit. From Eqs. (5.66) to (5.65), the total free energy functional is given by

5.2 Density Functional Approach

$$\mathcal{F}_{SPB}[\phi(\mathbf{r}), n_{\pm}(\mathbf{r})] = \int d\mathbf{r}[-\frac{\varepsilon}{8\pi}|\nabla\phi|^2 + zen_+\phi - zen_-\phi - \mu_+n_+ - \mu_-n_-] \quad (5.71)$$

$$+ \frac{-k_BT}{a^3}\int d\mathbf{r}[n_+a^3\ln(n_+a^3) + n_-a^3\ln(n_-a^3)$$
$$+ (1 - n_+a^3 - n_-a^3)\ln(1 - n_+a^3 - n_-a^3)] \quad (5.72)$$

To understand the effects of this novel entropic term, $\ln(1 - n_+a^3 - n_-a^3)$, consider a situation where the system is in contact with a highly charged surface. The counterions (say positive ions) will be attracted to the surface, thereby forming a dense layer. While ionic layering is favorable electrostatically, this is accompanied by a loss of solvent entropy as the solvent near the surface is expelled to make "room" for the ions. So as $n_+ \to 1/a^3$ in the layer, the solvent entropic cost is accounted for by the last term in $\mathcal{F}_{SPB}$. This motivated Borukhov et al. to call it the solvent entropy. Minimization of the free energy leads to a saturation of the density profile as seen in Fig. 5.2.

Finally, to obtain the SPB equation, a variational procedure with respect to $\phi(\mathbf{r})$ and $n_{\pm}(\mathbf{r})$ is performed (as was done in the last section), leading to the following Euler-Lagrange equations,

$$\frac{\delta F}{\delta \phi} = \frac{-\varepsilon}{4\pi}\nabla^2\phi + zen_+ - zen_- = 0 \quad (5.73)$$

$$\frac{\delta F}{\delta n_{\pm}} = \pm ze\phi - \mu_{\pm} + k_BT \ln \frac{n_{\pm}a^3}{1 - n_+a^3 - n_-a^3} = 0 \quad (5.74)$$

The first equation is just Poisson's equation. The second equation can be rewritten in this form,

$$\frac{n_{\pm}a^3}{1 - n_+a^3 - n_-a^3} = e^{\beta\mu_{\pm}}e^{\mp ze\beta\phi} \quad (5.75)$$

Given the definition of $\mu_{\pm}$ derived in the last section, these two equations are easily solvable for n_+ and n_- in terms of ϕ and n^b, to give Eq. 5.63. The latter, when combined with Poisson's equation, produces the SPB model (Eq. 5.62).

5.3 Generalized SPB Theory

The SPB model due to Borukhov, et al., considers ions to have a single size a. In this section, we generalize the model to treat ions with arbitrary size. This makes the theory applicable in a more general setting, where invariably, ions have different sizes, either due to their specific interactions with the solvent (hydration shells, etc ...) or to their chemical nature. A generalized SPB theory (GSPB), would allow us to tackle problems such as ion competition, where for example a smaller monovalent counterion exhibits more affinity to a charged surface than a larger divalent counterion.

A free energy approach would generate a functional that depends on all the ionic densities n_i given by $\mathcal{F}[\phi(\mathbf{r}), n_1(\mathbf{r}), \ldots, n_n(\mathbf{r})]$. As before, Poisson's equation comes from $\dfrac{\delta \mathcal{F}}{\delta \phi} = 0$, while $\dfrac{\delta \mathcal{F}}{\delta n_i} = 0$ generates non-algebraic equations for $n_1, \ldots, n_n$. Hence, an equation of motion in the form of Eq. 5.62 is not readily obtainable from this procedure. However, as mentioned earlier, one can still find the optimal n_i's directly from the free energy, using for example Monte-Carlo functional minimization. In order to derive a single differential equation that describes the generalized steric Poisson-Boltzmann, we need to derive the partition function of the system. Since the partition function will only be a function of ϕ and μ, effectively decoupling the n_i's from each other. The latter are easily obtained from the partition function in the usual way. To accomplish this, we generalize the procedure due to Chu et al. [14]. The grand-canonical ensemble partition function Ξ, can be derived by counting the possible occupancies of a lattice cell by the n ionic species. Consider a system with n ion types, each with diameter a_i. Suppose that the ion(s) with the largest diameter have a diameter a that sets the size of the lattice cell a^3. The cell volume occupied by an ion of diameter a_i is $v_i^3 a^3$, where $v_i = a_i/a$. The occupation possibilities of a lattice cell, where we only consider full occupation or empty, are as follows:

- Unoccupied cell:
$$\Xi_1 = 1. \tag{5.76}$$

- A cell occupied by a *single* ion of type i of volume a^3:
$$\Xi_2 = \sum_i{}' e^{-\beta(\mathrm{sgn}(i) z_i e \phi - \mu_i)}, \tag{5.77}$$

the prime indicates that the sum is limited to ions of diameter a.
- A cell occupied by m ions of types i that satisfy $\sum_i \alpha_i v_i = 1$, where α_i is the multiplicity of ions of type i. Note that the total number of ions in the cell is then given by $m = \sum_i \alpha_i$. This occupancy condition is satisfied trivially when all m ions are of the same type Q, i.e. $m = \alpha_Q = (v_Q)^{-1}$. The partition function of the general case is given by:

$$\Xi_3 = \sum{}'' \prod_i \left(\sum_{j=1}^{\alpha_i} \binom{\alpha_i}{j} (e^{-\beta(\mathrm{sgn}(i) z_i e \phi - \mu_i)})^j \right) \tag{5.78}$$

where the double prime indicates the sum over all possible ways to have m ions that satisfy $\sum_i \alpha_i v_i = 1$, and the product is over distinct ionic species i that make up the collection of m ions. The term in the parentheses counts the possible ways we can fill a cell with ions of type i with multiplicity $\alpha_i > 1$.

5.3 Generalized SPB Theory

The total partition function per lattice cell, is the sum of the individual partition functions,

$$\Xi[\phi(\mathbf{r})] = 1 + \sum_{i}^{\prime} e^{-\beta(\text{sgn}(i)z_i e\phi - \mu_i)} + \sum_{i}^{\prime\prime} \prod_{j=1}^{\alpha_i} \left(\sum_{j=1}^{\alpha_i} \binom{\alpha_i}{j} (e^{-\beta(\text{sgn}(i)z_i e\phi - \mu_i)})^j \right) \tag{5.79}$$

where the binomial coefficient is defined as $\binom{n}{p} = n!/p!(n-p)!$. In the case, where $n = 3$, with $v_1 = 1/2$ and $v_2 = v_3 = 1$ but $z_2 \neq z_3$, there is only one way to have more than one ion fully occupying a cell and that is with $m = 2$, with both ions of type 1, i.e their multiplicity is $\alpha_1 = 2$. Applying this to the partition function gives,

$$\Xi[\phi(\mathbf{r})] = 1 + e^{-\beta(\text{sgn}(2)z_2 e\phi(\mathbf{r}) - \mu_2)} + e^{-\beta(\text{sgn}(3)z_3 e\phi(\mathbf{r}) - \mu_3)}$$

$$+ \sum_{n=1}^{\alpha_1} \binom{\alpha_1}{n} \left(e^{-\beta(\text{sgn}(1)z_1 e\phi - \mu_1)} \right)^n \tag{5.80}$$

$$= (1 + e^{-\beta(\text{sgn}(1)z_1 e\phi - \mu_1)})^2 + e^{-\beta(\text{sgn}(2)z_2 e\phi(\mathbf{r}) - \mu_2)} + e^{-\beta(\text{sgn}(3)z_3 e\phi(\mathbf{r}) - \mu_3)} \tag{5.81}$$

The last line is obtained by combining the term that involves the sum and 1 to get the binomial expression. This expression is identical to the one derived in [14] [Eq. 1].

As in Sect. 5.1, the chemical potential of ion type i, is found by inverting

$$n_i^b = \frac{1}{Na^3} \frac{\partial \ln \Xi^N}{\partial (\beta \mu_i)} \bigg|_{\phi(\mathbf{r}) = 0} \tag{5.82}$$

where N is the number of lattice sites. While, $n_i(\mathbf{r})$ is given by,

$$n_i(\mathbf{r}) = \frac{1}{a^3 \Xi} \frac{\partial \Xi}{\partial (\beta \mu_i)} \tag{5.83}$$

The GSPB equation of motion is then given by

$$\nabla^2 \phi(\mathbf{r}) = -4\pi/\varepsilon \sum_i^n \text{sgn}(i) z_i e \frac{1}{a^3 \Xi} \frac{\partial \Xi}{\partial (\beta \mu_i)} \tag{5.84}$$

It is easy to check that in the case $n = 2$ ($v_1 = v_2 = 1$, $z_1 = z_2 = z$), 5.84 produces the SPB equation of motion (Eq. 5.62).

To illustrate the predictions of the GSPB model, we consider three case studies, all with $n = 3$ with 2 counter-ions and 1 co-ions. We consider the counter-ions, A$^+$ and B (counter-ion) with diameters $a_{A^+} = 1$ Å and $a_B = 10$ Å (or $a_B = 15$ Å) respectively. The co-ion Cl$^-$ has diameter $a_{Cl^-} = 3.6$ Å. Since, the largest counter-ion is B, then its size sets the lattice cell dimension, i.e $a_B^3 = a^3$. Then, we vary the valency

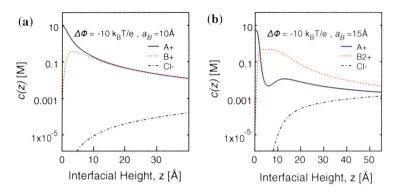

Fig. 5.3 Ion competition as predicted by the GSPB equation. All numerical solutions are obtained in a medium of dielectric constant $\varepsilon = 80$, $T = 300$ K. The electrolyte is made up of the following ions: A^+, B and Cl^-. The bulk ionic density is kept fixed at $c_{A^+}^b = c_B^b = 0.1$ M, and $c_{Cl^-}^b = 0.1\text{M} + z_B c_B^b$. An electrostatic potential difference of $\Delta\phi = -10 k_B T/e$ is applied. **a** Competition between two monovalent ions with different sizes. **b** Competition between a small monovalent ion and a larger divalent ion

of B ($z_B = 1$ or $z_B = 2$) to show the effects of volume-excluded interactions in conjunction with varying degrees of electrostatic attraction to the surface. Numerical solutions of Eq. 5.84 applied to this system are performed to find the ion distributions shown in Fig. 5.3. Figure 5.3a shows enhancement of A^+ and Ci^+ for $z > 10$ Å, since both ions are monovalent this enhancement at the surface produces an equal gain in electrostatic energy. However, closer to the surface the smaller A^+ is further enhanced at the interface, while the larger B^+ is depleted since the solvent entropy cost to layer A^+ is much smaller than the cost to layer the larger B^+. In Fig 5.3b, now a divalent B^{2+} can induce a bigger gain in electrostatic energy when enhanced at the interface, but due to large size 15 Å, at a much larger entropic cost. Hence, the GSPB theory predicts that a small *monovalent* ion (diameter of 1 Å) would outcompete a much larger divalent ion. While this result seems reasonable from energy arguments presented above, the GSPB theory is a mean-field theory whose results for divalent ions at such large surface potentials ($10 k_B T$) are suggestive at best. In Sect. 5.5, we will determine the range of validity of the SPB theory at a liquid/liquid interface as a function of surface electrostatic energy, $e\phi(0)$.

5.4 Numerical Implementation of Electrostatic Boundary Value Problems

The Poisson-Boltzmann equation and modifications of it presented in this chapter and others do not admit exact solutions in closed analytic form as a result of their nonlinearity. For application to the electrified liquid/liquid interface, we are interested in finding solutions to these models in a planar geometry, where in fact the

5.4 Numerical Implementation of Electrostatic Boundary Value Problems

Poisson-Boltzmann equation can be solved exactly only in the $z:z$ symmetric electrolyte if the surface charge σ is known, forming a well-posed initial value problem (IVP) (or equivalently, if one knows $\phi(0)$, where $z=0$ is the location of the interface, by using Gauss's law, the problem can be expressed as an IVP). On the other hand, the electrolyte/electrolyte interface on the other hand constitutes a well-formed boundary value problem (BVP). The applied potential across the system $\Delta\phi_{cell}$ is known from the electrochemical setup and using measured values of the potential of zero charge, $\Delta\phi = \phi_{\text{water}} - \phi_{\text{oil}}$, is accurately determined (see Chap. 3), where the subscripts water and oil refer to bulk water and bulk oil respectively. We use the following convention, the upper half-plane ($z>0$) is occupied by water and the lower half-plane ($z<0$) is occupied by oil (DCE). This implies that we know the value of the electrostatic potential, at two points, $\phi(\pm\infty)$. But the value of ϕ at $z=0$ is only constrained by the continuity of the electrostatic potential across the interface. In what follows, we present a numerical implementation of this BVP which produces a solution of the PB equation, and its variations, in the form of the electrostatic potential and the ion distributions.

We have used the computational suite Mathematica (Wolfram Research Inc, Version 8), to solve the second-order nonlinear differential equations of the electrical double layer models. Mathematica is a powerful computational package and programming language, with symbolic (exact) and numerical capabilities. It comes with an extensive library of compiled mathematical functions that covers the fields of statistics, pure and applied mathematics, data and image processing, etc...Mathematica solves differential equations (ordinary and partial) in exact form (when a solution exists) through the function DSolve[]. Numerical differential equation solving is handled by the NDSolve[] function which is approximately 1500 pages of optimized C-language code, containing dozens of standard differential equation solving methods such as "Runge-Kutta" and less conventional though powerful schemes such as the "Chasing" method. A book on numerical differential equation solving with Mathematica [15] and references therein has a thorough overview of the theory and code relevant to this functionality. We exclusively employ the "Shooting" method in conjunction with NDSolve[] in our code PBSolve[] (outlined in the appendix). To illustrate how the "Shooting" method works, we consider the example where we have an electrolyte in contact with a surface. In this 1-phase problem, assume a known potential drop given by ΔV. We want to solve a PB-like equation over the interval [0, z_{bulk}], the system is defined by:

$$\phi''(z) = S(\phi(z)); \; \phi(0) = \Delta V, \phi(z_{bulk}) = 0. \qquad (5.85)$$

where $''$ signifies second-order differentiation with respect to z, z_{bulk} is usually taken to be 10–15 Debye lengths away from the interface. When this system is given to NDSolve[] with the option **Method** → "Shooting" specified, the following is done (behind the scenes). The system is converted to an IVP,

$$\phi''(z) = S(\phi(z)); \; \phi(0) = \Delta V, \phi'(0) = \alpha. \qquad (5.86)$$

Now the system can be directly integrated (numerically) if α is known. Hence, the problem is reduced to determining α such that $\phi_\alpha(z_{bulk}) = 0$, a root finding problem (which is solved by FindRoot[]), where different α's are tried until α_{sol} is found that satisfies $\phi(z_{bulk}) = 0$. Given α_{sol}, the system defined by (5.86) is integrated. Returning the solution $\phi(z)$ as an interpolating function. The "Shooting" method is a very fast computational scheme, with solutions usually obtained in a few seconds. PBSolve[] does not use the internal error estimations of Mathematica, but has its own error measures accepting a solution returned by NDSolve[], only when it satisfies the boundary conditions within an error margin of at least 10^{-6} V (typical solutions have errors of 10^{-7} V). For 1-phase $z : z$ electrolyte, an exact analytic of the PB equation exists [16]. In Fig. 5.4 we compare the latter, to the numerical solution of this system obtained by PBSolve[]. We have computed electrostatic potentials for a 1:1, 2:2 and 3:3 electrolyte. The figures shows excellent agreement between the two solutions.

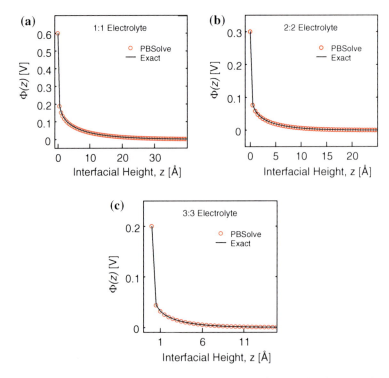

Fig. 5.4 Comparison between numerical and analytic solution of PB equation for a $z : z$ electrolyte. All solutions are obtained in a medium of dielectric constant $\varepsilon = 80$, $T = 300$ K, with the bulk concentration of $n^b = 0.1$ M. **a** 1:1 electrolyte, $\Delta\phi = 0.6$ V. **b** 2:2 electrolyte, $\Delta\phi = 0.3$ V. **c** 3:3 electrolyte, $\Delta\phi = 0.2$ V

5.4 Numerical Implementation of Electrostatic Boundary Value Problems

For further comparison, we plot $\phi(z)$ on a log-scale in Fig. 5.5a and the difference between the analytic and numeric solutions in Fig. 5.5b. As advertised, typical errors in the boundary conditions are on the order of 10^{-7} V.

The drawback of the "Shooting" method is its robustness. For instance, $\Delta\phi$'s used in Fig. 5.4 are the highest potentials at which solutions can be obtained from PBSolve[] within the prescribed errors. This is due to the fact that the Poisson-Boltzmann equation is highly singular near $z = 0$, hence the "Shooting" method procedure has difficulty finding a root that satisfies the Boundary value equation, and becomes highly sensitive to the starting point of this search. However, one should note that long before those potentials are reached, PB has ceased to predict physical results for the ion distribution. For instance, for a 3 : 3 electrolyte at $\Delta\phi = 0.2$ V, PB predicts that $c(0) \approx 10^9$ M !

For models less singular than PB (e.g SPB), PBSolve[] can find solutions at much larger potential drops. The solutions of PBSolve for the SPB models were compared to numerical solutions obtained in [10], and agree very closely. PBSolve[] was also tested against the "Quasi-Linearisation" scheme that converts the second order differential equation into an integro-differential equation for the electric field, and was used by Luo et al. in [17] to solve the PB equation, with the two methods being in close agreement as shown in Fig. 5.6.

For a 2-phase system, with permittivities $\varepsilon_{1,2}$, PBSolve[] works as follows:

- Define BVP1$[\phi_1(z), \alpha] = \{\phi_1''(z) = S_1(\phi_1(z)); \phi_1(0) = \alpha - \Delta V, \phi'(z_{\text{bulk phase 1}}) = 0\}$ in phase 1.
- Define BVP2$[\phi_2(z), \alpha] = \{\phi_2''(z) = S_2(\phi_2(z)); \phi_2(0) = \alpha, \phi'(z_{\text{bulk phase 2}}) = 0\}$ in phase 2.
- NDSolve[] is setup to solve BVP1 and BVP2 once α is assigned.
- FindRoot[] finds α_{sol}, that is α that satisfies $\varepsilon_1 \phi'_{\alpha,1}(0) = \varepsilon_2 \phi'_{\alpha,2}(0)$.
- Given α_{sol}, NDSolve[BVP1$[\phi_1(z), \alpha_{sol}]$, BVP2$[\phi_2(z), \alpha_{sol}]$] solves for $\phi(z)$.

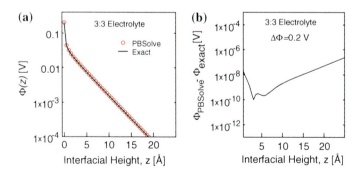

Fig. 5.5 Errors in the numerical solution of PB equation for a 3 : 3 electrolyte with $\Delta\phi = 0.2$ V. **a** Same plot as Fig. 5.4c on a log scale. **b** Difference between the exact and numerical solutions at each point of the solution interval

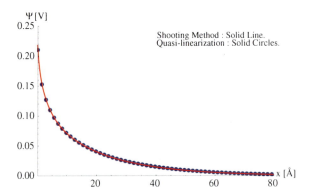

Fig. 5.6 Comparison of the numerical solution of PB equation for a 1 : 1 electrolyte using the Shooting Method and the "Quasi-Linearisation" scheme

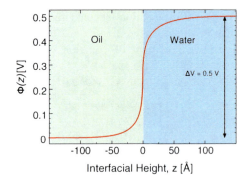

Fig. 5.7 Numerical solution of the electrostatic potential (PB equation) in 1 : 1/1 : 1 electrolyte with $\Delta\phi = 0.5$ V. Solution obtained using PBSolve[] under the following conditions: $\varepsilon_1 = 78.95$ (water), $\varepsilon_2 = 10.43$ (DCE), $T = 296$ K, $n_1^b = 0.01$ M, and $n_2^b = 0.0027$ M

A solution is typically found in less than 15 s on a 2.4 GHz Intel Core 2 Duo. The entire code is very compact (a few lines), efficient due to its use of native Mathematica functions (NDSolve[], FindRoot[]), and is included in the appendix. Shown in Fig. 5.7 is a numerical solution of the PB equation for a water/oil electrolyte system in Chap. 3 at an applied potential of 0.5 V. Note that all electrostatics (potentials, density profiles, etc ...) are computed using PBSolve[] in this chapter and others, unless noted otherwise.

5.5 Experimental Tests of the SPB Theory

The experimental system under study is the liquid/liquid interface between two immiscible electrolyte solutions: a 10 mM aqueous solution of LiCl and a 5 mM solution of bis(triphenyl phosphoranylidene) ammonium tetrakis(pentafluorophenyl)

5.5 Experimental Tests of the SPB Theory

borate (BTPPA$^+$, TPFB$^-$) in 1,2-dichloroethane (DCE). This system only differs from the electrolyte/electrolyte system described in Chap. 3, in using LiCl at a smaller concentration as the aqueous electrolyte instead of NaCl. Hence, all electrochemical studies (cyclic voltammetry, surface tension measurements ...) presented in Chap. 3 are qualitatively similar for the two systems (see [18] for a detailed electrochemical study of the 10 mM LiCl system).

As previously discussed, the electric potential difference between the water and oil (DCE) phases, $\Delta\phi$, is given by the applied potential difference across the electrochemical cell minus the potential of zero charge, $\Delta\phi = \Delta\phi_{\text{cell}}^{\text{w-o}} - \Delta\phi_{\text{PZC}}^{\text{w-o}}$ where $\Delta\phi_{\text{PZC}}^{\text{w-o}} = 365 \pm 4$ mV as determined by a measurement of the surface excess charge. Figure 5.8 illustrates the surface excess charge σ on either side of the liquid/liquid interface as determined by measurements of the interfacial tension. The latter is fitted to a second-order spline interpolation, the Lippmann equation, Eq. 3.7, was then used to determine the surface excess charge. As we charge the system, enhancement of ions on either side of the interface, produces an excess interfacial charge. Since, the system is globally neutral, we must have $|\sigma_o| = |\sigma_w| \equiv \sigma$, where as usual, o: oil and w: water.

The excess interfacial charge can also be directly computed from the ion density profiles, as follows

$$\sigma_o = e \int_{-\infty}^{0} dz (n^+(z) - n^-(z)) \tag{5.87}$$

for the oil phase, n^+ represents the ion density of BTPPA$^+$, and n^- that of TPFB$^-$. The excess interfacial charge in the water phase is similarly defined, with n^+ being the ion density of Li$^+$, and n^- that of Cl$^-$, with the integration now extending from

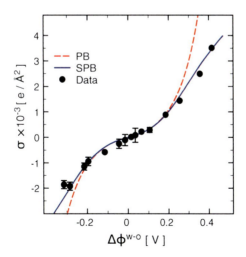

Fig. 5.8 Surface excess charge of the liquid/liquid interface as a function of electric potential difference, at T = 296 K, in units of electron charge e per Å^2

the interface ($z = 0$) to the aqueous bulk ($+\infty$). The very large Gibbs energies of transfer of the ions between the water and DCE phases allows us to ignore ion partitioning between the phases and to approximate the bulk ion concentrations, n_b as the dissociated concentrations (2.689 mM) of BTPPA$^+$ and TPFB$^-$ in DCE and initial concentrations of Li$^+$ and Cl$^-$ in water. The ion density profiles are computed from the PB equation,

$$\frac{d^2\phi(z)}{dz^2} = \frac{8\pi e n_b}{\varepsilon} \sinh(e\beta\phi(z)) \tag{5.88}$$

where ε is the dielectric constant of water (=78.95), when (5.88) is applied to the aqueous electrolyte, or the dielectric constant of DCE (=10.43), when applied to the oil phase. In Sect. 5.4, we outlined the numerical methods used to solve this type of electrostatic problems, where we imposed the following boundary conditions (Figs. 5.9 and 5.10):

- Discontinuity of the electric field at the interface: $\varepsilon_w E(z)|_{z=0^+} = \varepsilon_o E(z)|_{z=0^-}$.
- Continuity of the electrostatic potential at the interface: $\phi(z)|_{z=0^+} = \phi(z)|_{z=0^-}$.
- Bulk charge neutrality: $\lim_{z\to+\infty} E(z) = 0$, $\lim_{z\to-\infty} E(z) = 0$.
- Difference in potential between the bulks, equals the measured interfacial potential: $\Delta\phi = \Delta V$.

Note that the above boundary conditions are imposed in *all* applications of double-layers models to the electrified liquid/liquid interface presented in this work. Similar to the PB model, we solve the SPB equation (5.62), projected on the normal to the interface:

$$\frac{d^2\phi(z)}{dz^2} = \frac{8\pi e n_b}{\varepsilon} \frac{\sinh(z e \beta\phi(z))}{1 - 2a^3 n_b + 2a^3 n_b \cosh(z e \beta\phi(z))} \tag{5.89}$$

The size a in (5.89) is taken to be the size of the ion whose concentration is enhanced at the interface. For example, when $\Delta\phi > 0$, a in the water phase is the size of Li$^+$

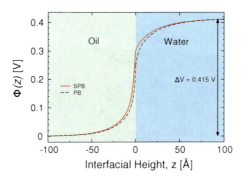

Fig. 5.9 Electrostatic potential distribution predicted from the PB and SPB models at an interfacial potential of 0.415 V

5.5 Experimental Tests of the SPB Theory

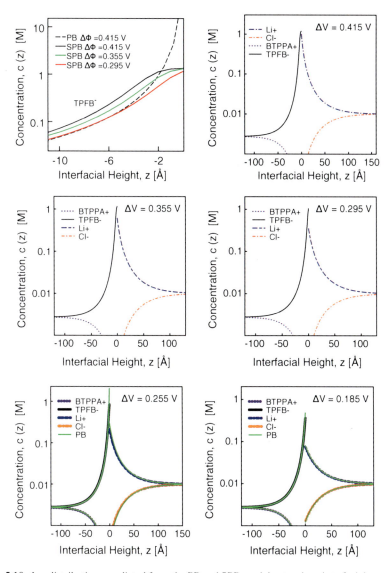

Fig. 5.10 Ion distributions predicted from the PB and SPB models at various interfacial potentials. Potentials in the range of 0.415 to 0.295V only show SPB density profiles

and a in the DCE phase is the size of TPFB$^-$. This is justified by the fact that at those potentials where steric effects are strong the depleted ion has an interfacial concentration that is orders of magnitude smaller than that of the enhanced ion. The ion concentration profiles predicted by SPB take the form of two back-to-back double layers at the interface Fig. 6.8, in qualitative agreement with Poisson-Boltzmann theory. Starting at $\Delta\phi = 0.295$ V, the concentration profile predicted from (5.89) is bound by $1/a^3$ and hence does not reach the unphysical values predicted by PB. The parameter a, in (5.89) is an effective ion size that we determine by fitting to the surface excess charge data. This is a natural consequence of the fact that our theory is a mean-field approximation. Note that the effective size is a result of the interactions between ions and is distinct from the hydrated ion size. A thorough discussion of this subtle difference, pertinent to double layer models, was presented in [8].

The solid line in Fig. 5.8 is computed from SPB by numerically integrating the calculated concentration profiles, the same was done for the PB equation. For $\Delta\phi < 0$, the ion size a in the water or DCE phase is set to be the van der Waals (vdW) diameter of 3.6 Å for Cl^- [19] or 11 Å for $BTPPA^+$ [18], respectively. For $\Delta\phi > 0$, the size $a_w = 1.44$ Å is the vdW diameter of Li^+, while $a_o = 11.2 \pm 0.1$ Å of $TPFB^-$ in the DCE phase is the only fitting parameter and is determined by fitting the SPB theory to the surface excess charge data. The fitting routine uses PBSolve to solve the SPB equation at different potentials given some size parameter. The calculated surface excess charge for a fixed ion size is then compared to the data by calculating the χ^2 value. This is done for $8\ \text{Å} < a_o < 13\ \text{Å}$. The ionic size given the best fit $a_o = 11.2 \pm 0.1$ Å is slightly larger than the calculated vdW diameter of 10.0 Å for $TPFB^-$ [18], as might be expected for a weakly solvated large ion. Varying the effective sizes of the other ions from their vdW diameters does not yield appreciably better fits to the data, which indicates, for example, that hydration of Li^+ does not play a significant role in determining the excess charge when $TPFB^-$ is on the DCE side of the interface. As discussed at the beginning of the chapter, the main difference between the SPB and PB model is the saturation of the density profile at the close-packing limit. This assumption produces a surface excess charge that is bounded from above, producing a good fit to the surface excess charge data, over a larger potential range than the Poisson-Botlzmann equation, as shown in Fig. 5.8. However, in the region where $\Delta\phi > 0.3$ V, the SPB surface excess charge does not follow (qualitatively) the trend of the data. Comparison of the reflectivity predicted by SPB to the x-ray data allows for more definitive conclusions on the range of validity of the theory.

Following the methodology of Chap. 4, we simulate the reflectivity from the SPB model with the interfacial roughness as our only fitting parameter. From Table 5.1,

Table 5.1 The potential dependent interfacial roughness of the 10 mM LiCl system: capillary wave theory and fits to the X-ray data

	Capillary wave theory (Å)	SPB (± 0.20 Å)	PB[a] (± 0.20 Å)
$\Delta\phi = 0.415$ V	5.25	4.88	5.25
$\Delta\phi = 0.355$ V	4.92	4.48	4.92
$\Delta\phi = 0.295$ V	4.79	4.42	4.79
$\Delta\phi = 0.255$ V	4.53	4.20	4.36
$\Delta\phi = 0.185$ V	4.52	4.12	4.28

[a] 0.415 V, 0.355 V, and 0.295 V, roughness values are not fitted for PB model. The fitted roughness at these potentials is more than 3 standard deviations

5.5 Experimental Tests of the SPB Theory

the fitted values of the potential-dependent roughness agree with values calculated from capillary wave theory using the measured interfacial tension within two standard deviations, though the fitted values are consistently smaller than the calculated values. For comparison, the dashed lines in Fig. 5.11 illustrate fits using PB. In the range of $0\,\text{V} < \Delta\phi \leq 0.255\,\text{V}$, SPB and PB yield similar reflectivity curves in agreement with the data. This reflects the fact that small ionic interfacial densities are present at these low potentials, leading to negligible steric effects. In addition, broadening of the density profiles with the ionic sizes and the interfacial roughness, produces electron density profiles in close resemblance for the two models. When $\Delta\phi = 0.295\,\text{V}$, clear deviations between PB and the data are seen, with SPB in excellent agreement, as shown in Fig. 5.12. At $\Delta\phi = 0.355, 0.415\,\text{V}$, the PB equation greatly overestimates $R(Q_z)/R_F(Q_z)$ primarily because it predicts unphysical ion concentrations of TPFB$^-$ that exceed the close-packing density of $1/a^3$ at the interface, while the agreement between $R(Q_z)/R_F(Q_z)$ and SPB is modest, with the latter predicting interfacial condensation of TPFB$^-$ with an average layer density of $1.7\,\text{nm}^2$ per ion at the highest potential.

The comparison between the theoretical models and the x-ray data are consistent with the results drawn from electrocapillary measurements. SPB captures the essential features of the potential dependent reflectivity data, however, it is important to note that the higher Q_z values of R/R_F in Fig. 5.12 are underestimated at higher potentials ($\Delta\phi \geq 0.355\,\text{V}$). This indicates that while the inclusion of solvent entropy in (5.89) leads to a stabilization of the ionic condensation, the SPB equation provides, at best, a first approximation to the functional form of the ion concentration

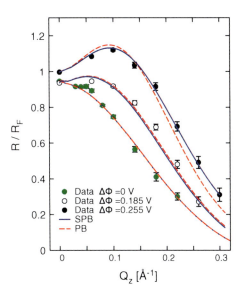

Fig. 5.11 X-ray $R(Q_z)/R_F(Q_z)$ compared to SPB and PB predictions in the potential range 0–0.255 V. For clarity, the data and simulated curves for 0 and 0.185 V are offset by -0.1

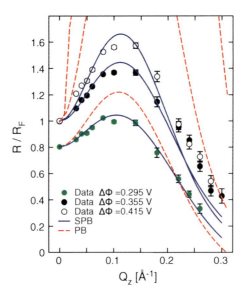

Fig. 5.12 X-ray $R(Q_z)/R_F(Q_z)$ compared to SPB and PB predictions in the potential range 0.295–0.415 V. For clarity, the data and simulated curves for 0.295 V are offset by -0.2

profile. This shortcoming of SPB was not apparent from a comparison to only the surface excess charge data shown in Fig. 5.8. This analysis demonstrates the need to use structural measurements, and not just thermodynamic data, to test models of ion distributions at soft-matter interfaces. The discrepancy between the SPB model and the data at the higher potentials can be attributed to the combination of large interfacial electrostatic energy ($\geq 14 k_B T$) and a large Bjerrum length in DCE ($\ell_B \approx 54$ Å), which should generate strong electrostatic ion-ion correlations for TPFB$^-$ at the interface even though it is a monovalent ion. These correlations are not included in the mean-field approximations of the sterically modified Poisson-Boltzmann equation and will be discussed in Chap. 7. However, the above analysis demonstrates the need to accurately account for ion-solvent correlations. This is treated in the next chapter using molecular dynamics simulations.

5.6 Concluding Remarks

In summary, we have presented a theoretical treatment of solvent correlations in a double-layer model, at the mean-field level, based on a lattice Coulomb gas formalism. The resultant theory is in the form of a modified Poisson-Boltzmann equation. We generalized the SPB model beyond its original formulation by Borukhov et al. to account for arbitrary ionic sizes. Using a numerical code devised to compute the electrostatics of boundary value problems, the ion density profiles of the SPB model

5.6 Concluding Remarks

were calculated as function of interfacial potential. The predictions of this sterically modified PB were tested against x-ray reflectivity data and electrocapillary measurements at the electrified liquid/liquid interface, establishing its utility in describing steric effects in ion distributions near a moderately charged liquid/liquid interface and demonstrating its advantages over the Poisson-Boltzmann equation in that same setting.

References

1. Stern, O.: Z. Elekt. Angew. Phys. Chem. **30**, 508 (1924)
2. Kjellander, R., kesson, T.A., Jönsson, B., Marčelja, S.: Double layer interactions in mono- and divalent electrolytes: A comparison of the anisotropic hnc theory and monte carlo simulations. J. Chem. Phys. **97**(2), 1424–1431 (1992)
3. Kjellander, R., Marcelja, S.: Double-layer interaction in the primitive model and the corresponding Poisson-Boltzmann description. J. Phys. Chem. **90**(7), 1230–1232 (1986)
4. Ninham, B.W., Parsegian, V.A.: Electrostatic potential between surfaces bearing ionizable groups in ionic equilibrium with physiologic saline solution. J. Theor. Biol. **31**(3), 405–428 (1971)
5. Attard, P., Mitchell, D.J., Ninham, B.W.: Beyond Poisson-Boltzmann: images and correlations in the electric double layer. i. counterions only. J. Chem. Phys. **88**(8), 4987–4996 (1988)
6. Attard, P., Mitchell, D.J., Ninham, B.W.: Beyond Poisson-Boltzmann: images and correlations in the electric double layer. ii. symmetric electrolyte. J. Chem. Phys. **89**(7), 4358–4367 (1988)
7. Antypov, D., Barbosa, M., Holm, C.: Incorporation of excluded-volume correlations into poisson-boltzmann theory. Phys. Rev. E **71**(6), 061106 (2005)
8. Wernersson, E., Kjellander, R., Lyklema, J.: Charge inversion and ion ion correlation effects at the mercury/aqueous mgso4 interface: toward the solution of a long-standing issue. J. Phys. Chem. C **114**, 1849–1866 (2010)
9. Borukhov, I., Andelman, D, Orland, H.: Adsorption of large ions from an electrolyte solution: a modified Poisson-Boltzmann equation. Electrochim. Acta **46**, 221–229 (2000)
10. Borukhov, I., Andelman, D., Orland, H.: Steric effects in electrolytes: a modified Poisson-Boltzmann equation. Phys. Rev. Lett. **79**, 435 (1997)
11. Stratonovich, R.L.: On a method of calculating quantum distribution functions. Sov. Phys. Doklady **2**, 416 (1958)
12. Peskin, M.E., Schroeder, D.V.: Introduction to Quantum Field Theory. Addison-Wesley Advanced Book Program, Reading, MA (1995)
13. Zinn-Justin, J.: Quantum Field Theory and Critical Phenomena. Oxford University Press, Oxford (2002)
14. Chu, V.B., Bai, Y., Lipfert, J., Herschlag, D., Doniach, S.: Evaluation of ion binding to dna duplexes using a size-modified Poisson-Boltzmann theory. Biophys. J. **93**(9), 3202–3209 (2007)
15. Sofroniu, M., Knapp, R.: Advanced Numerical Differential Equation Solving In Mathematica. Wolfram Mathematica Tutorial Collection. Wolfram Research, Inc., Champaign, IL (2008)
16. Schmickler, W.: Interfacial Electrochemistry. Oxford University Press, Oxford (1996)
17. Luo, G., Malkova, S., Yoon, J., Schultz, D.G., Lin, B., Meron, M., Benjamin, I., Vanysek, P., Schlossman, M.L.: Ion distributions near a liquid-liquid interface. Science **311**, 216–218 (2006)
18. Hou, B.: Ion Distributions at Electrified Liquid-Liquid Interfaces: An agreement between X-ray reflectivity analysis and macroscopic measurements. Doctoral dissertation, University of Illinois at Chicago (2011)
19. Marcus, Y.: Ionic radii in aqueous solutions. Chem. Rev. **88**(8), 1475–1498 (1988)

Chapter 6
Molecular Dynamics Simulation of Solvent Correlations

One of the main challenges encountered in the investigations of the electrical double layer is the complexity of ion specific effects, especially in regard to quantifying ion-solvent correlations. This complexity is further accentuated at interfaces, where ion solvation is inherently connected to the inhomogeneities in the solvent structure. Often, theories based on statistical mechanics of particles and continuum coarse-grained models offer simple and somewhat qualitatively correct pictures of solvation in the bulk [1, 2]. Nevertheless, the predictive power of these models is restricted by the large number of chemical components normally present in a solution of practical relevance. Most importantly, ion-solvent correlations in an inhomogeneous system, e.g. liquid/liquid interface, can only be described properly if the underlying interfacial structure is also accounted for accurately. This by itself is a daunting task, where most of the progress has been in the description of simple fluids near surfaces [3]. The advent of computer simulations has permitted comparisons between the molecular descriptions of the latter to the analytical models. Furthermore, simulations have offered a significantly novel and direct methodology to compare molecular models with experimental results, and in some instances, even facilitating the interpretation of the latter. The main difficulty in carrying out computer simulations, specifically classical molecular dynamics (MD), which was solely used in this work, resides almost entirely in determining the intermolecular and intramolecular potentials to define the system's potential energy surface. These potentials, commonly coined force fields, are often determined from ab initio quantum computations or fitted against empirical properties of the system at hand. Henceforth, force fields are very much system dependent and only occasionally transferrable. As outlined in Chap. 1, we will use MD simulations to map out the free energy profile of an ion near the liquid/liquid interface. This is interpreted as the contribution of ion-solvent correlations to the free energy of the electrical double layer.

The chapter is outlined as follows. We introduce in Sect. 6.1 specifics to the MD simulation of a water/DCE interface. In Sect. 6.2, we discuss the methodology used to simulate the solvent potential of mean force (PMF) for a number of atomic ions. The last section is devoted to determining the force fields of $TPFB^-$ and to the simulation of its solvent PMF.

6.1 MD Simulation of a Liquid/Liquid Interface

One of the first MD simulations of a neat liquid/liquid interface was performed by Benjamin [4]. The picture that emerged from this work and subsequently corroborated by others [5], is that of a liquid/liquid interface that is molecularly sharp undergoing wavelike distortions, bearing close resemblance to the thermal excitations predicted by capillary wave theory. All the MD simulations reported in this work were performed using a custom MD code (C-language), courtesy of Dr. Ilan Benjamin. The water potential used is the flexible SPC (simple point charge) model [6]. SPC is a three-site interaction model, where each hydrogen and oxygen carry a partial charge interacting through a Coulomb potential to model the electrostatic interaction, and oxygen atoms interact through a Lennard-Jones 6–12 potential that serves to model the van der Waals dispersion forces and repulsions due to electron–electron interactions, that are both quantum mechanical in nature. The SPC model reliably reproduces many of the empirical properties of water such as the density and the dielectric constant, yet does not include many-body interactions such as polarizability. In fact, none of the force fields we used are polarizable, since no consensus has formed in the MD simulation literature or community on the proper way to account for these many-body effects. Furthermore, the modeling of water is a very active research field, with few force fields able to withstand the scrutiny of comparison against ever more precise experimental studies [7, 8]. Dang and co-workers used polarizable force fields for both water and DCE molecules, with their simulations producing an interfacial structure similar to the one obtained by the use of the non-polarizable force fields parametrized by Dr. Benjamin and used here. The non-bonded interactions for both water and DCE molecules are represented by

$$U_{\text{NB}} = \sum_{i,j} \left(\frac{q_i q_j}{r_{ij}} + 4\epsilon_{ij} \left[\left(\frac{\sigma_{ij}}{r_{ij}} \right)^{12} - \left(\frac{\sigma_{ij}}{r_{ij}} \right)^{6} \right] \right) \quad (6.1)$$

with i indexing atoms ($i \neq j$), and the atomic Lennard-Jones (LJ) parameters presented in Table 6.1. The Lorentz-Berthelot mixing rules are employed to calculate interactions between atoms i and j,

$$\sigma_{ij} = \frac{\sigma_i + \sigma_j}{2}, \quad \epsilon_{ij} = \sqrt{\epsilon_i \epsilon_j} \quad (6.2)$$

Table 6.1 Partial charges and Lennard-Jones parameters for water and DCE models

Atom	σ (Å)	ϵ (kcal/mol)	$q(e)$
CH_2	3.98	0.114	0.227
Cl	3.16	0.5	−0.227
O	3.17	0.155	−0.82
H	2.81	0.017	0.41

6.1 MD Simulation of a Liquid/Liquid Interface

The DCE force field is represented by a four-site interaction model with the CH$_2$ groups of the molecule treated as a united atom, due to Benjamin [4]. The intramolecular potential energy is given by

$$U_B = \frac{k_{CC}}{2}(r_{CC} - r_{CC}^{eq})^2 + \frac{k_{CCl}}{2}(r_{CCl1} - r_{CCl}^{eq})^2 + \frac{k_{CCl}}{2}(r_{CCl2} - r_{CCl}^{eq})^2 \quad (6.3)$$

$$+ \frac{k_\theta}{2}(\theta_1 - \theta_{CCCl}^{eq})^2 + \frac{k_\theta}{2}(\theta_2 - \theta_{CCCl}^{eq})^2 \quad (6.4)$$

$$+ \frac{V_1}{2}(1 + \cos\phi) + \frac{V_2}{2}(1 - \cos 2\phi) + \frac{V_3}{2}(1 + \cos 3\phi) \quad (6.5)$$

where the first and second lines express bond stretching and bending using a harmonic potential, where k is a constant and $\theta_{1,2}$ are bend angles $\widehat{CCCl}$. The third line represents torsion around the C–C bond using a cosine expansion in the torsional angle ϕ. The parameters that enter V_B (listed in Table 6.2) are determined by fitting to the empirical data (gauche/trans ratio and vibrational frequencies).

In the simulation of atomic ions (Sect. 6.2), the water/DCE system was modeled by a rectangular box in the xy plane, with dimensions 24.8×24.8 Å, the oil phase extends from $z = 0$ to $z \approx 40$ Å, while the aqueous phase reaches $z \approx -26$ Å. The dimensions of the liquid phases are determined from the observed bulk densities for a system with a total of 500 water molecules and 215 DCE molecules. In the simulation of TPFB$^-$, its large size required the use of a box with the following dimensions: 50×50 Å, with the same length of the liquid phases, as above, giving a total of 2,424 water molecules and 844 DCE molecules (see Fig. 6.1). The intermolecular interactions are smoothly cutoff at distances larger than half the length of the simulation box. The truncation of the long-range behavior of the Coulomb potential is treated using the reaction field correction [9]. The type of modeling discussed, gives rise to one liquid/liquid interface and two liquid/vapor interfaces, one at each z-edge of the simulation box. Benjamin has showed that these two liquid/vapor interfaces negligibly disturb the bulk liquids and consequently the liquid/liquid interface. All the simulations discussed in this chapter were done using the microcanonical ensemble (constant N, V, E), using a velocity Verlet algorithm to integrate the equations

Table 6.2 Parameters of bonded interactions of DCE

k_{CC}	620 kcal/mol Å^{-2}
r_{CC}^{eq}	1.53 Å
k_{CCl}	464 kcal/mol Å^{-2}
r_{Cl}^{eq}	1.787 Å
k_θ	88 kcal/mol rad^{-2}
θ_{CCCl}^{eq}	108.2°
V_1	−0.24 kcal/mol
V_2	0.1 kcal/mol
V_3	3.228 kcal/mol

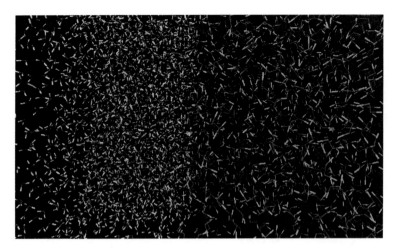

Fig. 6.1 Snapshot of the molecular dynamics simulation of the water/1,2-dichloroethane interface in the presence of an ion

of motion [9]. The water/DCE system is usually equilibrated for a few hundred picoseconds to attain a fixed temperature, using Nosé constant temperature dynamics and velocity rescaling.

6.2 Potential of Mean Force of Na$^+$, Li$^+$, and Cl$^-$

After obtaining an equilibrated liquid/liquid interface, an ion is inserted by replacing some of the solvent molecules. Since, we only treat atomic ions in this section, the expression of the potential energy of the system is unchanged, with the sum in (6.1) running over the ion's index as well. The only force field parameters that need to be specified then are the 6–12 LJ parameters, listed in Table 6.3. We choose the ionic intermolecular potentials that were found in [10] to produce hydration free energies and first hydration shell structure in reasonable agreement with experimental values.

Two simulation procedures were used depending on the definition of the PMF employed. In the case of Na$^+$, we used the overlapping windows method used previously by Benjamin to simulate adsorption energy of inorganic ions near the vapor/water interface [11], whereby the ion is constrained by a potential in the

Table 6.3 Lennard-Jones parameters for the simulated ions

Ion	σ (Å)	ϵ (kcal/mol)
Na$^+$	2.275	0.115
Li$^+$	1.594	0.133
Cl$^-$	3.934	0.832

6.2 Potential of Mean Force of Na$^+$, Li$^+$, and Cl$^-$

z-direction over a range of 3 Å to sample the system's phase space in this spatial region alone. The procedure then is to setup multiple system configurations where in each configuration the center of the potential is at a different z, allowing for a 1 Å overlap between two subsequents potential windows. Hence, to simulate the PMF over a region of 40 Å (from bulk to bulk), 20 potential windows are needed. The advantage of this procedure is that now each configuration undergoes an equilibration run and production run, separately on a different processor, thereby generating a computational load that is colloquially known as *embarrassingly parallel*. After the ion is inserted into the potential window, an initial small integration time step on the order of 10^{-3} to 10^{-1} fs is usually needed to start out the equilibration process, since the position of the inserted ion can be too close to a solvent molecule, producing very large interaction energies. Afterwards, a time step of 1 fs is used during the remainder of the equilibration which lasts for 500 ps. After the end of equilibration, i.e. the temperature of the system has reached the desired value, typically 300 ± 3 K, a simulation run is started where in each potential window we collect sampling statistics for 2 ns, using a time step of 1 fs. The system's total energy is monitored after every 20 ps (at the end of a trajectory) to ensure that it is conserved, typically rms deviations of no more than 0.1 % are experienced during a simulation. The ion (one-particle) probability density per potential window is then computed from,

$$p(z) = \langle e^{-\beta U_{ion}} \rangle \tag{6.6}$$

where U_{ion} is the potential energy of the ion (Coulomb + 6–12 LJ) and $\langle . \rangle$ represents the ensemble average (average over all trajectories). From the sampling of $p(z)$, we readily obtain the free energy,

$$f_{\text{solvent}}(z) = -kT \ln p(z) \tag{6.7}$$

In Fig. 6.2, we plot the free energy for each window. Notice that as the ion approaches the boundary of the potential window, there is a spike in $-kT \ln p(z)$, this could be traced to the increase in the potential energy of the ion as it "feels" the constraining potential. The data points at the boundaries have to be eliminated, since these points are artifacts and do not represent a physically meaningful situation. Furthermore, note that for windows deep in the liquid bulk phases ($z < -17$ Å and $z > 15$ Å) the free energy is nearly constant compared to the large increases near the interface ($z = 0$).

The free energy profile across the interface is obtained by matching $-kT \ln p(z)$ in the region of overlap between every two windows and shifting the entire profile by a constant so that the free energy is zero in bulk water, as shown in Fig. 6.3. As expected for a strongly hydrated alkali ion, the free energy starts to increase as the ion starts to interact with the DCE molecules, while deeper in the bulks the free energy is constant, reflecting thermodynamic equilibrium. The most stringent test one can impose on a simulated PMF is a comparison of its bulk value difference to the Gibbs free energy of transfer. For the sodium ion, the experimental free energy of transfer is $23 \pm 2 k_B T$ (Table 3.1) and is in excellent agreement with the simulation result,

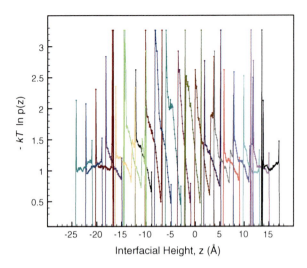

Fig. 6.2 MD simulated free energy of Na$^+$ in different potential windows across the water/DCE interface

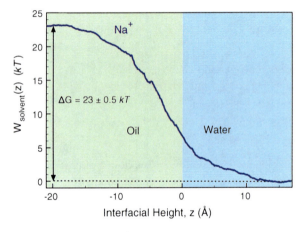

Fig. 6.3 MD simulated solvent PMF of Na$^+$. The small-scale structure in the PMF is due to random fluctuations in the window sampling as shown in 6.2

$23 \pm 0.5 k_B T$. Nevertheless, the comparison between the PMF bulk difference and the Gibbs free energy of transfer is not always straightforward as will be discussed later in the section.

In the free energy simulation of Cl$^-$, the overlapping window method did not produce a stable free energy profile that converged in bulk phases, as shown in Fig. 6.4. An inspection of the ensemble averaged solvent density profiles reveals that excessive mixing of the solvents occurs, greatly distorting the interface as shown Fig. 6.5, with the ion located 13 Å from the interface.

6.2 Potential of Mean Force of Na$^+$, Li$^+$, and Cl$^-$

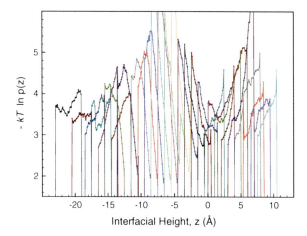

Fig. 6.4 MD simulated free energy of Cl$^-$ in different potential windows across the water/DCE interface

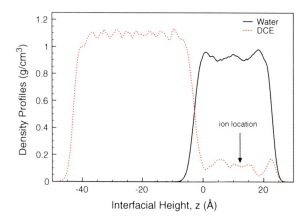

Fig. 6.5 Liquid/Liquid interface profile in the presence of Cl$^-$ constrained by a potential window

Several factors could be at play. First, the relatively large size of Cl$^-$ could be disturbing the interfacial structure and a larger simulation box needs to be used. Second, the interaction between Cl$^-$ and a DCE molecule may not be properly parametrized, recall that this interaction was obtained using the mixing rules, while ideally one would like to derive the force field parameters from empirical data and/or *ab initio* computations of a chloride ion in a dichloroethane solvent that are unavailable at the moment. Third, the constant motion of chloride in the potential window may be causing the solvent mixing. We did not test the first hypothesis, since a larger simulation box (four times bigger) is computationally expensive due to the N^2 scaling behavior of the MD code, where N is the number of particles. Furthermore, we will use a large simulation box when we simulate the PMF of TPFB$^-$. Also as mentioned, due to the lack of empirical and *ab initio* input we could not further test the Cl$^-$-DCE interaction parameters. To investigate the effect of the constraining potential on the simulation, we will use a different methodology to simulate the ionic free energy profile. This technique is different from the overlapping potential windows used up to this point, in that instead of constraining the ion to move over a small region, we completely freeze its spatial motion. The equilibration and production runs proceed as described earlier. The definition used for the potential of mean force is given by,

$$W_{\text{solvent}}(z) = -\int dz' \langle F(z - z') \rangle \qquad (6.8)$$

where $\langle F(z) \rangle$ is the z—component of the total force acting on the center of mass of the ion due to all $N - 1$ particles in the system, averaged over all trajectories. Due to its computational simplicity, we exclusively used this method in all simulations presented in the remainder of this chapter. In Fig. 6.6, $\langle F(z) \rangle$ is plotted for a Cl$^-$ simulation where each data point was sampled for 2 ns, but the force is far from converged in either bulk. Longer simulation times would have no effect on the force convergence, since a monitoring of the interaction energy of chloride with the solvent shows that the latter typically converges in less than 50 ps, as shown in Fig. 6.7. This indicates that the interface is still significantly disturbed by Cl$^-$ either due to its large size or less than optimal interactions with DCE.

To further test the origin of this instability, we use a soft restoring potential $(= -\alpha z^3)$ that a water molecule would become subject to if its height is less than 0, stopping it from further going into the DCE phase. Consequently this restoring force avoids the mixing of the two solvents as illustrated in Fig. 6.8, where a series of simulations were performed as a function of the strength of the restoring potential.

At the highest interaction strength, $\alpha = 40kT$, the interface is practically a hard-wall, with the solvent density profiles exhibiting oscillations characteristic of fluids layering near a hard-wall [12]. At lower interaction strengths, $\alpha = 0.2kT, 0.4kT$, the

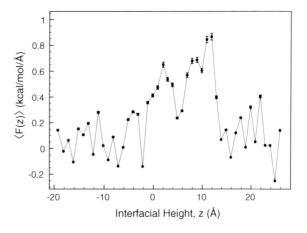

Fig. 6.6 Ensemble averaged force acting on chloride at different positions along the normal to the interface

6.2 Potential of Mean Force of Na$^+$, Li$^+$, and Cl$^-$

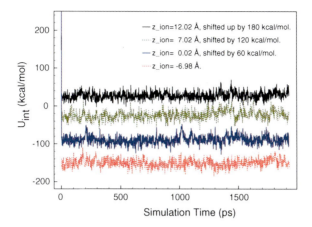

Fig. 6.7 Interaction energy of water and Cl$^-$ at different positions along the normal to the interface, monitored over the simulation run

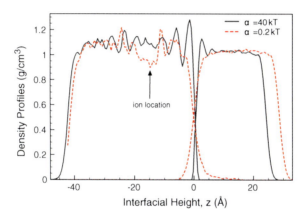

Fig. 6.8 Solvent density profiles at different interaction strengths of the water restoring potential

solvent density profiles resemble those of a stable water/DCE interface [4]. Further decrease in the interaction strength, $\alpha < 0.2kT$, produced density profiles resembling those in Fig. 6.5. The PMF simulation procedure outlined above was performed at these same restoring potentials, with the resultant free energy profiles shown in Fig. 6.9.

The PMF obtained are well-behaved and converge to a constant value in both bulk phases. The simulated free energies of transfer are at odds with the experimental value, 12.66 ± 0.8 kcal/mol (Table 3.1) with $\Delta W_{\text{solvent}} \approx 16$ kcal/mol with minimal variation between the three PMFs. The most noticeable difference between the three cases is the sharp increase in the free energy at $z \approx -5$ Å but diminishes in magnitude as α becomes smaller. We can conclude that the overall structure of the potential of mean force has been converged by tuning the strength of the restoring potential.

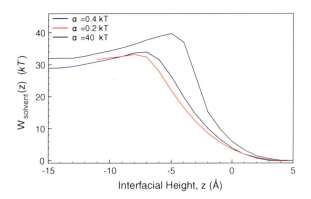

Fig. 6.9 Cl⁻ PMF using different interaction strengths of the water restoring potential

Using thermodynamic integration of a hydrated ion in water and in DCE, Rose and Benjamin [13] showed that the free energy of transfer is highly dependent on the number of water molecules within a cluster for small hydrophilic ions such as Li⁺ (see below). For larger ions such as Cl⁻, there is less of a dependence. Nevertheless, the correspondence between the Gibbs free energy of transfer and the PMF in the case of Cl⁻ is ambiguous due to the disturbance that ion transfer impinges on the interface. In addition, we can infer from the MD results of [13], which used the same force field parameters for the solvents and the ion as in this work, to produce Gibbs energies of transfer in good agreement with the data, that the Cl⁻-DCE interaction are well parametrized, at least in the bulk. A PMF simulation of Cl⁻ by Wick and Dang [14], using polarizable potentials for the solvents and the ion, produced results that structurally resembled ours but with the free energy difference in agreement with the experiments, and was used in our calculations of ion density profiles (see Chap. 4).

We have also carried out simulations of the PMF of Li⁺, which showed no disturbance of the interface, similar to results of Na⁺, further justifying our claim that the large mixing of the solvents is probably correlated with the size of the ion being transferred across the interface. The Li⁺ PMF simulation proceeded as described earlier with a close monitoring of the interaction energy between the ion and water to determine force convergence as shown in Fig. 6.7. To determine the effect of hydration shells on the free energy profile, we have performed two simulations, one with a restoring potential ($\alpha = 40kT$) and one without. The results shown in Fig. 6.10 are in agreement with the simulated free energies of transfer in Fig. 6 of [13], where $\Delta G \approx 60$ kcal/mol for the transfer of a bare Li⁺ ion to be compared with $\Delta W_{\text{solvent}} = 64$ kcal/mol for $\alpha = 40kT$. In the case of no restoring potential, $\Delta W_{\text{solvent}} = 28$ kcal/mol, an inspection of the radial distribution function $g(r)$ when Li⁺ is located at $z = -12.6$ Å (Fig. 6.11), a position at which $W_{\text{solvent}}(z)$ has almost reached its bulk value, reveals that the ion is solvated on average by approximately $1 - 2$ water molecules in its first hydration shell, a snapshot from the MD simulation

6.2 Potential of Mean Force of Na$^+$, Li$^+$, and Cl$^-$ 77

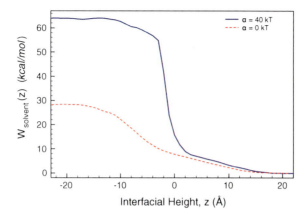

Fig. 6.10 Li$^+$ PMF at the water/DCE interface, with and without a restoring potential

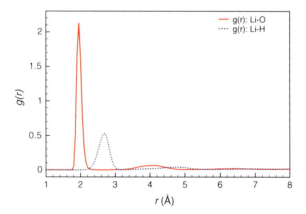

Fig. 6.11 Radial distribution functions of oxygen and hydrogen around a central Li$^+$ located at $z = -12.6$ Å in DCE

also confirms this picture Fig. 6.12. For a cluster that contains 2 water molecules + Li$^+$, $\Delta G \approx 30$ kcal/mol from [13], in excellent agreement with the PMF value.

6.3 Potential of Mean Force of TPFB$^-$

Force field parameters of TPFB$^-$ consist of defining the Lennard-Jones parameters (ϵ, σ) for each constituent atom (C,F,B), partial charges for Coulomb interactions, and intramolecular interactions. Since, the crystal structure of TPFB$^-$ is known, equilibrium bond properties that enter the force field are determined by averaging over the crystal structure values (r_{BC}^{eq}, r_{CF}^{eq}, r_{CC}^{eq}, θ_{CBC}^{eq}, θ_{BCC}^{eq}, θ_{CCF}^{eq} in Table 6.4). The harmonic force constants that define bond stretching and bending, involving combinations of

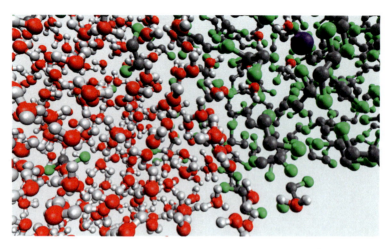

Fig. 6.12 Snapshot of the MD simulation with the ion located at $z = -12.6$ Å in DCE. For clarity, the size of the ion (*purple*) was magnified

C and F atoms (k_{CF}, k_{CC}, k_{CCF}, k_{CCC} in Table 6.4) are readily available in force field libraries such as the general Amber force field [15], and were used as is. No force field parameters involving boron were found though, due to its uncommon use in simulation of organic and biological compounds. Hence, the bond stretching constant of the C–B bond (k_{BC}) was determined from a vibrational analysis using Gaussian03 [16], the quantum model employed was second-order Moller-Plesset perturbation theory (MP2) in the aug-cc-pVDZ basis, a description of the computational details is given below. Since, boron enters in all bonded interactions that define the geometry of TPFB$^-$ in the simulation, its corresponding parameters (k_{CBC} and k_{CBC}) were set to preserve the equilibrium tetrahedral structure. Consequently, the bending constant of the C–B–C angle was tentatively set to be approximately equal to that of its counterpart for the C–C–C, with the equilibrium angle, $\widehat{CBC}$, set to be that of a tetrahedral, given that the crystal structure average ($\widehat{CBC} = 107.55°$) is very close to that geometry. The torsion potentials, $V_{F-C-C-C}$, are chosen to keep the fluorine atoms in the plane of the flurophenyl rings and $V_{C-C-B-C}$ to maintain the tetrahedral equilibrium configuration, as illustrated in a snapshot from the simulation (Fig. 6.13). In Table 6.4, we summarize the parameters that enter the intramolecular potentials, in the quadratic potential approximation (6.5), to model bonded interactions of TPFB$^-$.

The intramolecular potentials of TPFB$^-$ have a small contribution to the overall ion interaction energy compared to the Coulomb and Lennard-Jones interactions. Hence, accurate parametrization of the latter is necessary, especially for the C and F atoms. The Lennard-Jones parameters for aromatic carbon and fluorine were taken from the amber force field [17], while those of boron are taken from the literature [18], shown in Table 6.5. Several LJ parameters for C and F were tested, with the ones from the Amber force field producing the best agreement with the experimental free energy of transfer, while varying (ϵ_B, σ_B) did not affect the results of the simulation.

6.3 Potential of Mean Force of TPFB⁻

Table 6.4 Parameters of Bonded interactions of TPFB⁻

k_{BC}	631.74 kcal/mol Å^{-2}
r_{BC}^{eq}	1.66 Å
k_{CF}	368.70 kcal/mol Å^{-2}
r_{CF}^{eq}	1.35 Å
k_{CC}	589.70 kcal/mol Å^{-2}
r_{CC}^{eq}	1.38 Å
k_{CBC}	100.0 kcal/mol rad^{-2}
θ_{CBC}^{eq}	109.47°
k_{BCC}	200.0 kcal/mol rad^{-2}
θ_{BCC}^{eq}	113.4°
k_{CCF}	67.8 kcal/mol rad^{-2}
θ_{CCF}^{eq}	119.50°
k_{CCC}	69.8 kcal/mol rad^{-2}
θ_{CCF}^{eq}	120.0°
$V_{F-C-C-C}$	9 kcal/mol
$V_{C-C-B-C}$	90 kcal/mol

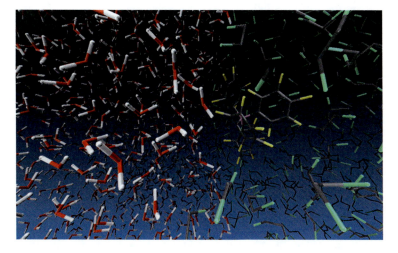

Fig. 6.13 Snapshot of the TPFB⁻ MD simulation

Table 6.5 Partial charges and Lennard-Jones parameters for TPFB⁻

Atom	σ (Å)	ϵ (kcal/mol)	q(e)
F	3.50	0.061	−0.145199
C	3.816	0.086	0.079332
B	3.543	0.095	1.2×10^{-5}

The atomic partial charges of TPFB⁻ were computed using the CHELPG algorithm implemented in the computational chemistry package NWCHEM [19]. The CHELPG algorithm consists of assigning a charge q_i to atom i in a molecule, then using a monopole expansion to calculate the resultant electrostatic potential (ESP),

$$V_{\text{ESP}}(\mathbf{r}) = \sum_{i}^{nuclei} \frac{q_i}{|\mathbf{r} - \mathbf{r}_i|} \tag{6.9}$$

V_{ESP} is then compared to the "true" molecular electrostatic potential, $V(\mathbf{r})$ defined by,

$$V(\mathbf{r}) = \sum_{i}^{nuclei} \frac{Z_i}{|\mathbf{r} - \mathbf{r}_i|} - \int d\mathbf{r}' \Psi(\mathbf{r}') \frac{1}{|\mathbf{r} - \mathbf{r}_i|} \Psi(\mathbf{r}') \tag{6.10}$$

where Z_i is the atomic number of nuclei i, and $\Psi(\mathbf{r}')$ is the electronic wavefunction of the molecule. The difference between $V_{\text{ESP}}(\mathbf{r})$ and $V(\mathbf{r})$ is minimized by fitting the partial charges q_i. This is done in the CHELPG algorithm by setting up a grid around the surface of the molecule and calculating the distribution of V_{ESP} [20] at these mesh points. A significant drawback of this method is the insensitivity of the electrostatic potential on the value of the partial charge of an atom that is far from the molecular surface. For instance, in the case of TPFB$^-$, a straightforward CHELPG computation of the partial charges gives a large charge on the boron of 2.62 e, which needs to be compensated by the carbons bonded to it, causing the latter to have an average charge of $-0.8\ e$ (see list in the appendix 7.4). Since such a distribution of charges would generate unrealistically high Coulomb interactions in the MD simulation, we choose to constrain the charge of the boron to zero. Also, to not overcomplicate the force fields, we constrain the carbon (fluorine) atoms to have the same charge. This is justified by the tetrahedral symmetry of the molecule and to force carbon atoms farther away from the van der Waals molecular surface to have reasonable charges. The partial charges obtained are listed in Table 6.5. The quantum computations were done using Moller-Plesset's second-order perturbation theory (MP2) in the Hartree-Fock theory (HF) [21]. HF theory ignores electron-electron correlation, thereby assuming that an electron interacts only with the electrostatic field due to the $N - 1$ electrons in the system, and gives an electronic energy equivalent to Moller-Plesset first-order perturbation. MP2 introduces electron-electron repulsions, that originate in the (full) Schrödinger Hamiltonian, as a perturbation. This scheme is found to give more accurate electrostatic potential distributions than HF, with the MP2 computed moments in good agreement with empirical data. Of crucial importance is the basis or atomic orbitals used to do the quantum calculations. We used the aug-cc-pVDZ basis [22], which are a large collection of Gaussian functions, producing 25 functions per atom, using the following polarization (4s3p2d) to ensure that bonding and electric field properties are properly described. The aug-cc-pVDZ basis sets represent the state the art in correlated quantum calculations and ensure that moments calculated from the CHELPG are at least in close agreement with empirical data, which in the case of TPFB$^-$ is lacking.

6.3 Potential of Mean Force of TPFB$^-$

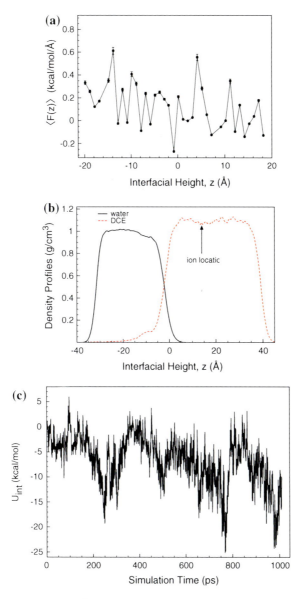

Fig. 6.14 TPFB$^-$ simulation at the water/DCE interface without a restoring potential. **a** Ensemble averaged force over 1.7 ns. **b** Averaged Interfacial density profile with ion located at $z = 13.7$ Å. **c** Interaction energy of TPFB$^-$ with water over 1 ns, $z_{ion} = 13.7$ Å

MP2/aug-cc-pVDZ produce a dipole moment of 0.015886 (a.u) and a quadrupole moment tensor ($Q_{xx} = -1.930683$, $Q_{yy} = 1.299866$, $Q_{zz} = 0.6301817$), given here as a future reference.

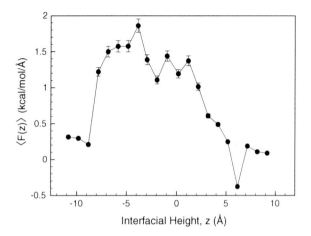

Fig. 6.15 TPFB$^-$ simulation of the average force acting on the center of mass of the ion with a restoring potential

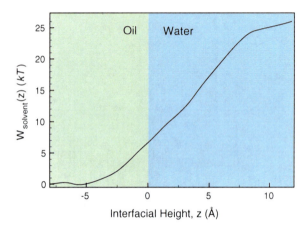

Fig. 6.16 TPFB$^-$ Solvent free energy profile simulated at the water/DCE interface

With the force field parameters defined by Tables 6.4 and 6.5, the PMF MD simulation proceeds as described earlier. The ion is inserted into the simulation box, 500 ps equilibration run ensues. Then the ion's center of mass is translated by 1 Å in the z-direction, at each new position a 500 ps equilibration run is performed with the solvents' molecules initial configuration given by the previously equilibrated system. This continues until we have a series of equilibrated boxes with ionic positions spanning bulk water to bulk DCE. Then at each fixed z, with the molecular ion's spatial degrees frozen, we collect sampling statistics for 2 ns. The ensemble averaged force (over the last 1.7 ns of the simulation), and solvents density profile presented in Fig. 6.14 are reminiscent of the simulation of Cl$^-$. Such a behavior is expected, in light of the discussion in Sect. 6.2. Even though a simulation box with four times

6.3 Potential of Mean Force of TPFB⁻

the surface area is used, we expect that a much larger TPFB⁻ with diameter of 10 Å would create significant distortions as seen in Fig. B. This is further confirmed by looking at the interaction energy of TPFB⁻-Water, in Fig. C, where the latter did not converge even over the span of 1 ns.

Unfortunately, due to the prohibitively expensive scaling of the simulation as N^2, and a 2 ns simulation time (per position) requiring 408 h of processor time, we were unable to further increase the size of the system. Instead, we use a restoring potential to maintain a relatively stable water/DCE interface during the PMF simulation. The strength of the restoring potential was tuned with the interfacial profile monitored, this produced a minimum interaction strength of $0.25kT$, below which excessive mixing occurs. It is worth noting that a simulation of TPFB⁻ in a smaller MD box like the one we used for the atomic ions, necessitated the use of a much higher restoring potential strength ($>1kT$) to avoid excessive solvent mixing. With the strength of the restoring potential fixed, a series of simulations are undertaken with the resultant average force as a function of distance along the interfacial normal shown in Fig. 6.15, exhibiting convergence in both bulks. The solvent PMF obtained from the force data is given in Fig. 6.16, where $\Delta W_{solvent} = 26.0 \pm 0.5kT$, is in agreement with the experimental value of $\Delta G = 29.3 \pm 2\ kT$, indicating that the MD PMF can predict properties of the TPFB⁻ at the water/DCE interface that are accurate. The issue of the PMF predicting correct ion-solvent correlations will be addressed in Chap. 7.

6.4 Concluding Remarks

The simulated solvent free energy profiles represent a convenient and compact method to account for ion-solvent correlations in a free energy formalism of the electrical double layer. In the next chapter, it will be shown that $f_{solvent}(z)$ is not only a necessary ingredient for a proper, and as it turns out accurate, description of ion distributions in the presence of strong electrostatic ion correlations, but is crucial to extract physically sensible results from ion-ion correlation models.

References

1. Marcus, Y.: Ion Solvation. Wiley, New York (1985)
2. Ben-Naim, A.: Molecular Theory of Solutions. Oxford University Press, Oxford (2006)
3. Hansen, J.-P., McDonald, I.: Theory of Simple Liquids, 3rd edn. Elsevier, Amsterdam (2006)
4. Benjamin, I.: Theoretical study of water/1,2-dichloroethane interface: Structure, dynamics and conformation equilibrium at the liquid-liquid interface. J. Chem. Phys. **97**, 1432 (1992)
5. Wick, C.D., Dang, L.X.: Molecular dynamics study of ion transfer and distribution at the interface of water and 1,2-dichloroethane. J. Phys. Chem. C **112**(3), 647–649 (2008)
6. Berendsen, H. J. C., Postma, J. P. M., van Gunstersen, W. F., and Hermans, J.: Intermolecular Forces. D. Reidel Publishing Company, Dordrecht (1981)

7. Schmidt, J.R., Roberts, S., Loparo, J., Tokmakoff, A., Fayer, M., Skinner, J.: Are water simulation models consistent with steady-state and ultrafast vibrational spectroscopy experiments? Chem. Phys. **341**, 143–157 (Jan 2007)
8. Zhang, Z., Piatkowski, L., Bakker, H.J., Bonn, M.: Communication: Interfacial water structure revealed by ultrafast two-dimensional surface vibrational spectroscopy. J. Chem. Phys. **135**(2), 021101 (2011)
9. Allen, M.P., Tildesley, D.J.: Computer Simulation of Liquids. Clarendon, Oxford (1987)
10. Benjamin, I.: Structure and dynamics of hydrated ions in a water-immiscible organic solvent. J. Phys. Chem. B **112**, 15801–15806 (2008)
11. Benjamin, I.: Theoretical study of ion solvation at the water liquid-vapor interface. J. Chem. Phys. **95**(5), 3698 (1991)
12. Henderson, D. (ed.): Fundamentals of Inhomogeneous Fluids. Marcel Dekker, New York (1992)
13. Rose, D., Benjamin, I.: Free energy of transfer of hydrated ion clusters from water to an immiscible organic solvent. J. Phys. Chem. B **113**, 9296–9303 (2009)
14. Wick, C.D., Dang, L.X.: Recent advances in understanding transfer ions across aqueous interfaces. Chem. Phys. Lett. **458**, 1–5 (2008)
15. Wang, J., Wolf, R.M., Caldwell, J.W., Kollman, P.A., Case, D.A.: Development and testing of a general amber force field. J. Comput. Chem. **25**, 1157–1174 (2004)
16. Frisch, M. J., Trucks, G. W., Schlegel, H. B., et al.: Gaussian 03. Gaussian, Inc. Wallingford (2004)
17. Cornell, W., Cieplak, P., Bayly, C.I., Gould, I.R., Merz, K.M., Ferguson, D.M., Spellmeyer, D.C., Fox, T., Caldwell, J.W., Kollman, P.A.: A second generation force field for the simulation of proteins, nucleic acids, and organic molecules. J. Am. Chem. Soc. **117**, 5179–5197 (1995)
18. Firlej, L., Kuchta, B., Wexler, C., Pfeifer, P.: Boron substituted graphene: energy landscape for hydrogen adsorption. Adsorption **15**, 312–317 (2009)
19. Valiev, M., Bylaska, E., Govind, N., Kowalski, K., Straatsma, T., van Dam, H., Wang, D., Nieplocha, J., Apra, E., Windus, T., de Jong, W.: Nwchem: a comprehensive and scalable open-source solution for large scale molecular simulations. Comput. Phys. Commun. **181**, 1477 (2010)
20. Cramer, C. J.: Essentials of Computational Chemistry : Theories and Models, 2nd edn. Wiley, New York (2004)
21. MØller, C., Plesset, M.S.: Note on an approximation treatment for many-electron systems. Phys. Rev. **46**(7), 618–622 (Oct 1934)
22. Dunning, T.H.: Gaussian basis sets for use in correlated molecular calculations. i. the atoms boron through neon and hydrogen. J. Chem. Phys **90**(2), 1007–1023 (1989)

Chapter 7
The Role of Electrostatic Ion Correlations in Ion Condensation

In the past decade, significant theoretical progress has been made in understanding the regime where electrostatic ion correlations dominate in a classical Coulomb fluid, i.e., the strong-coupling limit (SC). The SC models are often based on the picture of a 2-dimensional ionic layer near the charged surface due to Rouzina and Bloomfield [1]. These authors proposed that the character of electrostatic correlations amongst counter-ions near a charged surface is 2-dimensional rather than 3-dimensional. Hence, within this proposed picture, in the limit of zero temperature, the ground state of the counter-ions is given by the well-known Wigner crystal. Shklovskii and co-workers [2, 3] applied this idea to the study of correlations by assuming that the chemical potential of the Wigner crystal (extrapolated from MC simulations) describes the chemical potential of the condensed ionic layer even for non-zero temperature. These authors found that the diffuse tail of the density profile falls exponentially instead of an inverse power law as in the case of the Poisson-Boltzmann solution. Subsequent field-theoretic approaches to this $2d$ bound state were initiated by Netz and others [4, 5], giving qualitative agreement with Shklovskii's model, and offered a systematic derivation of the one-particle densities through a virial expansion in the inverse coupling constant $\Xi = 2\pi q^3 \ell_B^2 \sigma$, where q is the charge carried by the counter-ion, ℓ_B is the Bjerrum length, and σ is the surface charge density. The resultant expansion in Ξ is asymptotically exact and may offer the possibility of exact interpolation between the weak coupling limit of the PB theory and the SC regime. Excellent agreement is found between the SC field-theory and MC simulations as shown in the references above. Aside from the SC models discussed above, there is a long tradition of treating ion correlations using density functional theories with the correlation free energy extrapolated from the MC simulation of a one-component plasma developed by Stevens and Robbins [6] or from Debye-Hückel-like theories put forth by Penfold and coworkers [7] that are often based on the idea of a "correlation hole" to correct for the unphysical behavior of the DH density near the origin. The most popular model based on a correlation hole is that due to Nordholm [8] and is commonly called the Debye-Hückel-Hole theory of a one-component plasma, within which a correlation free energy can be exactly calculated using the process of Debye charging [7]. The density functional theories in Refs. [6, 7] are in the local density approximation (LDA). However, one-component plasma models are known to become unstable at large densities or large Bjerrum lengths. Hence, weighted density approximations (WDA) were used to avoid this instability [9–11]. Integral equations of liquid state theory were the first to show that like-charge attraction is due to ion correlations and are very powerful and accurate models [12–14] producing agreement with simulation and macroscopic data of surface forces [15] and surface excess charge [16]. However, a number of approximations enter the construction resulting in complicated theories that are known to produce "no solution" regions. All of the above models consider a structureless dielectric background, thereby neglecting the effects of solvent correlations. Hence, comparison of the theoretical results has been mostly limited to Monte Carlo simulations. The chemical specificity of ion interactions with a charged surface [17] and the complexity of solvent mediated effects [18] in a typical strongly correlated system, complicate the connection between theory and data. Therefore, an experimental analysis of SC models where ion–solvent interactions are adequately described is highly desirable. In this chapter we provide direct comparisons between the predictions of the DHH model, implemented using a weighted density approximation, and x-ray reflectivity measurements, as well as surface excess charge. In Sect. 7.1 we present a treatment of the DHH model. In Sect. 7.2 we use the latter in a WDA, in conjunction with solvent correlations simulated from MD simulation (see Chap. 6). In the last section, the prediction of this model is compared against the data.

7.1 The Debye-Hückel Hole Theory

The one-component plasma (OCP) model is a collection of N point charges with charge e in a rigid neutralizing continuum background with dielectric ϵ and volume V. We fix an ion at the origin $r = 0$ and derive how the remaining ions are distributed about it. The potential due to an ion at the origin, is given by the Poisson equation

$$\nabla^2 \phi = -\frac{4\pi}{\epsilon} \rho(r), \tag{7.1}$$

where ρ is the charge density. This system is one of the simplest Coulomb systems one can study characterized by only two length scales, the Bjerrum length $\ell_B = \beta e^2/\epsilon$, and the mean nearest-neighbor separation $d = 3/(4\pi n^b)^{1/3}$, where $n^b = N/V$ is the bulk density. These two length scales form a coupling constant $\Gamma = \ell_B/d$ that measures the strength of the interaction in the system. The charge density is given by

$$\rho(r) = e(n^b g(r) - n^b), \tag{7.2}$$

where the first term on the right hand side is the charge due to the ions and the second term is due to the negatively charged background. The average electrostatic potential

7.1 The Debye-Hückel Hole Theory

vanishes in this system. However, the inhomogeneity in the ion distribution around the central ion, given by the radial distribution $g(r)$ produces a nonzero electrostatic free energy. In fact, ignoring all interactions but the electrostatic one, $g(r)$ is given by

$$g(r) = e^{-\beta W(r)}, \qquad (7.3)$$

where $W(r)$ the potential of mean force, which we approximate by $e\phi(r)$, where as in Eq. (7.2), $\phi(r)$ is the electrostatic potential due to the central ion. Note that this approximation ignores electrostatic correlations due to the ionic atmosphere around the central ion. Since $\rho(r)$ is now defined we could insert it into Eq. (7.2) to obtain the Poisson-Boltzmann equation. Since no analytic solution exists for the latter in a spherical geometry, Debye and Hückel linearized the exponent in Eq. (7.3) to obtain an equation that only depends on $\phi(r)$ [19], $\rho(r) = -\beta e n^b \phi(r)$ putting this into Poisson's equation we find,

$$\nabla^2 \phi = \kappa^2 \phi(r), \qquad (7.4)$$

where $\kappa = (4\pi e^2 n^b \ell_B)^{1/2}$, is the inverse Debye screening length. The solution of the Helmholtz equation (7.4) gives the DH potential,

$$\phi(r) = e \frac{e^{-\kappa r}}{\epsilon r}. \qquad (7.5)$$

Consequently, the charge density is given by $\rho(r) = -\kappa^2 \exp[-\kappa r]/4\pi r$, and we immediately notice that while $\rho(r)$ should be bounded from below $\rho(0) \geq -en^b$, we have $\rho(r) \to -\infty$ as $r \to 0$. This pathology is easily traced to the linearization of the Boltzmann factor which is valid only for small deviations from the bulk density n^b. To correct this problematic behavior, Nordholm [8] postulated the existence of a correlation hole of radius s around the central ion, where co-ions are totally excluded. Note that in the DH theory of the OCP, the co-ions are being pushed away from the central ion, at large $\rho(r)$ this ultimately leads to negative $n(r)$, clearly an unphysical result. The correlation hole is then the byproduct of the strong repulsion between the co-ions surrounding the central ion. A repulsion that is energetically costly to overcome. In the region outside the hole, $r > s$, the linearization should still hold with the electrostatic potential now given by $C \frac{e^{-\kappa r}}{r}$, with C a constant that is determined from the continuity of $\rho(r)$ at $r = s$. Fixing the constant gives the following charge density distribution,

$$\rho(r) = \begin{cases} -en^b, & r < s, \\ -en^b s e^{-\kappa(r-s)}/r, & r \geq s, \end{cases} \qquad (7.6)$$

The radius of the correlation hole is found by imposing charge neutrality,

$$4\pi \int dr\, r^2 \rho(r) = -e, \qquad (7.7)$$

producing,

$$s = \frac{1}{\kappa}(\omega - 1),$$
$$\omega = (1 + 3\ell_B\kappa)^{1/3}. \tag{7.8}$$

The dependence of the correlation hole on Γ given by the expression in (7.8) conforms to the expectation of the interplay between kinetic and potential energy. For instance, in the limit $\Gamma \to \infty, s \to d$, with Γ representing an inverse temperature we expect that the internal energy of an ion is composed almost entirely of electrostatic energy with the ions strongly correlated and equivalently generating an exclusion region around the central ion, of radius equal to the average ion–ion distance. In the opposite limit of high temperature, $\Gamma \to 0, s \to \ell_B$ and the electrostatic repulsion is comparable to the thermal energy. Given the charge density, the electrostatic potential is easily solved and can be found elsewhere [20], with the potential due to the ionic atmosphere given by $\psi = -k_BT/2e(\omega^2 - 1)$. The free energy per particle, f^{DHH} due to correlations in the OCP can be calculated exactly by Debye charging the co-ions,

$$\beta f^{\text{DHH}} = \beta e \int_0^1 \psi(\lambda e) d\lambda \tag{7.9}$$
$$= \frac{1}{4}\left(1 + \frac{2\pi}{3\sqrt{3}} + \ln\left(\frac{\omega^2 + \omega + 1}{3}\right) - \omega^2 - \frac{2}{\sqrt{3}}\tan^{-1}\left(\frac{2\omega + 1}{\sqrt{3}}\right)\right) \tag{7.10}$$

This free energy per particle could then be used in a density functional theory to find the optimum density profiles. An important observation with a serious consequence for the density functional approach and the thermodynamics of the OCP plasma is the asymptotic behavior of f^{DHH}. In the limit of large densities, equivalently large Γ, the $-\omega^2$ term dominates over the other terms in Eq. (7.10). The free energy density then behaves as $-n\omega^2$, i.e, $nf^{\text{DHH}} \sim -n^{4/3}$ clearly not a convex function, yet meta-stable results are found if $n(z)$ does not cross some critical density as described in [11]. Nonetheless, it can be shown that this non-convexity can lead to negative pressures [21]. This thermodynamic instability is not restricted to the DHH model alone but is a general feature of a one-component plasma (e.g. from MC simulations) with the pressure becoming negative at $\Gamma > 3$ leading to a collapse, because a decrease in the volume is accompanied by a decrease in the electrostatic correlation free energy, and the entropy can no longer compensate for this inward force. This negative pressure is ultimately caused by ignoring the pressure due to the background. Levin in [20] proposed a way to deal with this instability that depends on the physical system at hand. In our application of the DHH theory of a OCP to a liquid/liquid interface this problem is bypassed by accounting for solvent correlations.

7.2 A Density Functional Theory of Ion Correlations

To describe ion correlations in the double layer of the 10 mM NaCl (H_2O)/5 mM BTPPATPFB (DCE) system, we use a density functional theory developed by Penfold et al. [7], and Stevens and Robbins [6], but amended here to account for MD simulated solvent correlations [22, 23]. We consider the grand potential per unit area to be composed of an ideal (Poisson-Boltzmann) contribution $\mathcal{F}^{PB}$ and an excess free energy due to correlations,

$$\frac{\beta}{A}\Omega[n_+, n_-] = \sum_{i=+,-}\left(\frac{\beta}{A}\Omega^{PB}[n_i] + \beta\int f_i^{sol}(z)n_i(z)dz\right) + \frac{\beta}{A}\mathcal{F}^{ion}[n_-(z)], \tag{7.11}$$

where $\Omega^{PB}[n_\pm]$ is the grand potential due to the ($\pm$) ion given in Eq. (2.1), consisting of the ideal gas entropy, the electrostatic interaction, and the chemical potentials. Note that for $z > 0$ or $z < 0$, (+) denotes Na^+ or $BTPPA^+$ and (−) denotes Cl^- or $TPFB^-$. To describe the *specific* interactions of an ion with the solvent, we have introduced the free energy profile per ion i, $f_i^{sol}(z)$. These were MD simulated for Na^+ and $TPFB^-$, and presented in Chap. 6. Since we are mostly interested in potentials where $BTPPA^+$ and Cl^- are largely depleted from the interfacial. The solvent free energy profile of Cl^- was MD simulated by Wick and Dang [24] and is used here as is, while f^{sol} of $BTPPA^+$ was fitted to interfacial excess charge assuming some functional form that mimics the MD simulated profiles [25]. We only include ion correlations for $TPFB^-$ due to the following considerations. Owing to the small dielectric constant of the oil phase $\epsilon^o = 10.43$, there is a strong electrostatic interaction between monovalent $TPFB^-$ ions, characterized by the coupling strength $\Gamma \approx 4$, where $d \approx 1.5$ nm is the mean nearest-neighbor separation at the interface for the highest probed potential ($\Delta\phi = 0.406$ V), and the Bjerrum length $\ell_B = \beta e^2/\epsilon^o \approx 5.5$ nm in DCE at $T = 296$ K. The correlations of $BTPPA^+$ are not included, as mentioned above, its interfacial density is orders of magnitude less than that of $TPFB^-$ (for instance, see Fig. 4.6). Moreover, due to the dilute bulk density in the oil phase, $n^b = 10^{-6}$ ions/$Å^3$, the strength of the interaction is weak with $\Gamma = 0.89$. In the aqueous phase ion correlations are negligible, the Bjerrum length is much smaller $\ell_B \approx 0.7$ nm, hence $\Gamma \ll 1$ by virtue of the presence of monovalent ions. In fact the Bjerrum length of monovalent ions in DCE is approximately equal to that of *trivalent* ions in water!

Minimization of Eq. (7.11) proceeds as before (see Chap. 2), ignoring ion correlations for the moment ($\mathcal{F}^{ion} = 0$),

$$\frac{\beta}{A}\frac{\delta\Omega[n_{\pm}]}{\delta n_{\pm}} = \frac{\beta}{A}\frac{\delta\Omega^{\rm PB}[n_{\pm}]}{\delta n_{\pm}} + \beta \int f_{\pm}^{sol}(z)\frac{\delta n_{\pm}(z)}{\delta n_{\pm}(z')}{\rm d}z$$
$$= \ln(\Lambda^3 n_{\pm}) + \beta\left(\pm e\phi(z) - \mu_{\pm} + f_{\pm}^{sol}(z)\right). \quad (7.12)$$

Solving for $n_{\pm}$, we find that the optimal density profile of Na^+, Cl^-, and $BTPPA^+$ is given by

$$n_{\pm}(z) = n^b \exp\left(\beta\left(\mp e\phi(z) - f_{\pm}^{sol}\right)\right). \quad (7.13)$$

To find the density profile of TPFB$^-$ including ion correlations, we use the Debye-Hückel-Hole theory (DHH) of the *homogeneous* OCP derived in the previous section. To account for the inhomogeneity in the system using a density functional theory, one can use the local density approximation (LDA) $n^b \to n_-(z)$ in Eq. (7.10) to obtain $f^{\rm DHH}(n_-(z))$. However for large densities or large Bjerrum lengths there is an instability in the computation of $n_-(z)$ within the LDA, known as a "structuring catastrophe" [7]. The origin of this instability is due to the asymptotic behavior of the correlation free energy discussed at the end of the last section. Specifically, during the computation of $n_-(z)$ within the LDA an increase in the local density is accompanied by a decrease in the excess chemical potential, which in turn increases the local density even more, this proceeds *ad infinitum* and in the words of Barbosa et al. "the overall system collapses to a point." [11], who also derived a critical density as a function of ion valency in water, beyond which the instability takes place. In our computations of the density profiles at the electrified liquid/liquid interface, the structuring catastrophe occurs at potentials higher than 180 mV, when the TPFB$^-$ density reaches a value of 0.5 M. To circumvent this, Groot [9] suggested using the DHH correlation free energy within a weighted density approximation (WDA) [26, 27], i.e., $f^{\rm DHH}(\bar{n}_-(z))$, where $\bar{n}_-(z)$ is the weighted density defined by,

$$\bar{n}_-(z) = \int n_-(z')w\left(|z - z'|, n_-(z)\right){\rm d}z', \quad (7.14)$$

where the weight function w depends on the local density and smoothes out any sharp fluctuations in the latter when calculating the excess chemical potential. Note that the inverse Debye length is now also a function of the weighted density, as well as the quantities it enters, i.e, s, and ω (see Eq. (7.8)). An expression for the weight function was originally derived in [9] based on consistency with the particle correlation functions of the homogeneous system. By comparison of the second variation of the free energy of the inhomogeneous DHH plasma to the direct correlation function, Groot arrived at the following w-function [9],

$$w(r) = \frac{3}{2\pi s^2}\left(\frac{1}{r} - \frac{1}{s}\right)\Theta(s - r) \quad (7.15)$$

with s the radius of the correlation hole defined in Eq. (7.8). Since Eq. (7.15) depends on the radial coordinate r, we need to project it on the z-axis to apply it in a planar

7.2 A Density Functional Theory of Ion Correlations

geometry. Applying the definition in Eq. (7.14) to $\bar{n}(r)$ and using the polar coordinates (ρ, ϕ)

$$\bar{n}_-(z) = \int dz' n_-(z') \int_0^{2\pi} d\phi \int_0^\infty \rho \, d\rho \; w(\sqrt{\rho^2 + |z-z'|}; n(z))$$

$$= 2\pi \int dz' n_-(z') \int_0^\infty \rho \, d\rho \frac{3}{2\pi s^2} \left(\frac{1}{\sqrt{\rho^2 + |z-z'|^2}} - \frac{1}{s} \right) \Theta(s - \sqrt{\rho^2 + |z-z'|^2})$$

$$= \frac{3}{s^2} \int dz' n_-(z') \int_0^{\sqrt{s^2 - |z-z'|^2}} d\rho\rho \left(\frac{1}{\sqrt{\rho^2 + |z-z'|^2}} - \frac{1}{s} \right)$$

$$= \frac{3}{s^2} \int dz' n_-(z') \left(\sqrt{\rho^2 + |z-z'|^2} - \frac{\rho^2}{2s} \right) \Bigg|_{\rho=0}^{\rho=\sqrt{s^2-|z-z'|^2}}$$

$$= \frac{3}{2s^3} \int dz' n_-(z') (s - |z-z'|)^2. \tag{7.16}$$

The same expression was arrived at in [10] (Eq. (29)) using a quadrature. The weighted density now defined, we can derive the density profile of TPFB$^-$ from Eq. (7.11) by minimization with respect to the *local* density n_-, with the correlation free energy given by

$$\mathcal{F}^{\mathrm{ion}}[n_-(z)] = \int n_-(z) f^{\mathrm{DHH}}(\bar{n}_-(z)) dz, \tag{7.17}$$

we find,

$$n_-(z) = n^b \exp[\beta(e\phi(z) - f_-^{sol}(z) - \mu_-^{ion}(z))],$$

$$\mu_-^{ion}(z) = \frac{\delta \mathcal{F}^{\mathrm{ion}}}{\delta n_-(z)}$$

$$= f^{\mathrm{DHH}}(n_-) + \int dz n_-(z) \frac{\delta f^{\mathrm{DHH}}(\bar{n}_-)}{\delta n_-}. \tag{7.18}$$

Since the excess chemical potential due to ion correlations, μ^{ion} depends on n_- through Eq. (7.14), to calculate the ion distributions, Poisson's equation and Eqs. (7.13), (7.16), and (7.18) are numerically solved to self-consistency and define what we call the PB/MD/DHH model. The solution to Poisson's equation, the electrostatic potential $\phi(z)$, satisfies the same boundary conditions used in p. xxx. The algorithm used to self-consistently solve for the density profiles for the NaCl/BTPPATPFB system from the PB/MD/DHH model is the following:

- Choose $n_-^{\mathrm{guess}}(z)$, for $z < 0$.
- Find $s(n_-^{\mathrm{guess}}(z))$ from (7.8).
- Compute $\bar{n}_-(z)$ from (7.16) with $n_-(z) \to n_-^{\mathrm{guess}}(z)$.
- Find $f^{\mathrm{DHH}}(\bar{n}_-(z))$.
- Compute μ_{ex} from (7.18).

- Define Poisson's equation with $n_-(z)$ given by (7.18) and $n_+(z)$ given by (7.13) for $z < 0$ and $n_\pm$ defined by (7.13) when $z > 0$.
- Solve Poisson's equation using PBSolve.
- The solution of Poisson's equation $\phi(z)$, is used to find $n_-(z)^{sol}$, again using (7.18).
- If $max\{|n_+(z)^{sol} - n_+^{guess}(z)|\}$ < Tolerance, then the computation has converged.
- If $max\{|n_+(z)^{sol} - n_+^{guess}(z)|\}$ > Tolerance, then $n_+^{new\ guess}(z) \equiv n_+(z)^{sol}$ and the loop is repeated.

A few notes about the above algorithm (code included in the appendix): (a) For fast convergence $n_-^{guess}(z)$ at a potential $\Delta\phi^1$ can be taken to be the density profile at that same potential from (7.13) that defines the PB/MD model where no ion correlations are present, (b) the code scans a list of differences between the guess and the solution at each z, after the max value is found it is compared to the convergence criterion or the tolerance. We choose a tolerance of 10^{-6}M, this produces an approximate error of 0.1 % in n^b used in the computation, and less than 10^{-4} % in the maximum value of $n_-(z)$. In the next section, we compare the density profiles to the x-ray reflectivity measurements and the interfacial excess charge.

7.3 Experimental Tests of the PB/MD/DHH Theory

In this section, we compare the ion density profiles predicted by the models discussed in the last section, to the excess surface charge data and XR measurements of the 10 mM NaCl/5 mM BTPPATPFB. The electrochemistry of this system is similar to the 100 mM NaCl system presented in Chap. 3. Interfacial measurements determined a potential of zero charge of 0.374 V [25] with respect to which, all interfacial potentials $\Delta\phi$ discussed are referenced. In Fig. 7.1, the ion concentration profiles are illustrated. The PB/MD model that only includes f^{sol} for each ion and ignores electrostatic ion correlations predicts TPFB$^-$ density profiles with a relatively long diffuse tail, with a peak concentrations increasing with increasing potential. The high free energy cost for TPFB$^-$ to approach the interface due to unfavorable interactions with the water as shown in Fig. 6.16, the ion density peaks at about $a/2$, where $a = 1$ nm is the ion's van der Waals diameter and decreases rapidly past that point. With ion correlations added to this model as described in the last chapter, the PB/MD/DHH model predicts TPFB$^-$ ion density profiles qualitatively and quantitatively different than the PB/MD model, characterized by a fast falling diffuse tail in qualitative agreement with other SC models such as Shklovskii's Wigner Crystal model [3]. Much higher densities are predicted by PB/MD/DHH by comparison with the PB/MD model, reflecting the basic fact that electrostatic ion correlations produce a system configuration that is energetically lower than a configuration where those correlations are not included. This translates into more ions condensing onto the polarized interface.

7.3 Experimental Tests of the PB/MD/DHH Theory

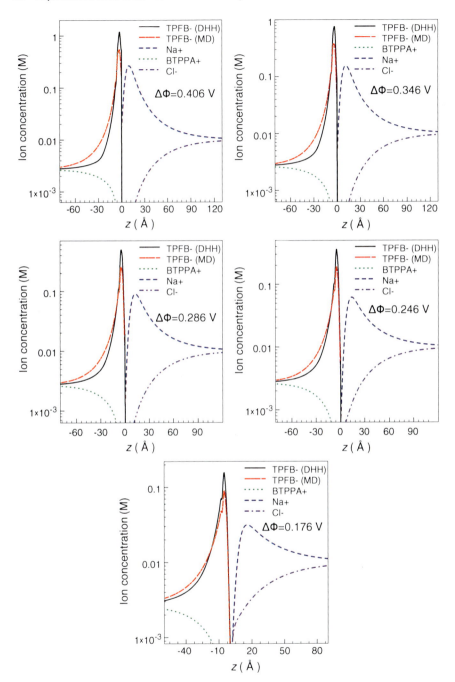

Fig. 7.1 Ion concentration profiles predicted from the PB/MD/DHH (DHH) model and the PB/MD (MD) model at all interfacial potentials probed

Fig. 7.2 TPFB⁻ reduced density profiles predicted from the PB/MD/DHH model as function of interfacial height at different potentials and coupling constants. The ion diameter $a = 10$ Å

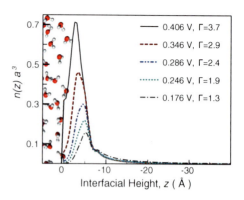

Progressively higher densities of TPFB⁻, as illustrated in Fig. 7.2 are predicted as systematic polarization of the interface increases the coupling constant Γ from a weakly correlated system $\Gamma \approx 1$ to a strongly correlated one with $\Gamma \approx 4$ at the highest potential $\Delta\phi = 0.406$ V. However, the reduced density never quite reaches 1 or close packing indicating that the system is not close to the regime where overcharging (i.e enhancement of BTPPA⁺) would start to occur. Ion correlation energies reach approximately $4\,k_B T$ with the functional form of the excess chemical potential also reflecting the energetics due to solvent correlations. Since, the highest value of the electric field is at $z = 0$, one would expect that $\mu^{ion}(z)$ would have a minimum at that z, yet unfavorable interactions with water forces the TPFB⁻ layer to get as close to the interface as possible gaining electrostatic energy before solvent correlations become too penalizing (Fig. 7.3).

We first test the predictions of PB/MD/DHH against the interfacial excess charge measurements (see Chap. 3 for an outline on how these measurements are performed). As described in Sect. 5.5, we integrate the theoretical density profiles to find the interfacial charge σ. This thermodynamic measurement serves to constrain global electrostatic properties of the theoretical models. Poisson-Boltzmann theory predicts $\sigma(\phi)$

Fig. 7.3 Excess chemical potentials due to ion correlations as a function of interfacial height at different potentials

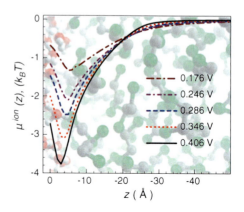

7.3 Experimental Tests of the PB/MD/DHH Theory

Fig. 7.4 10 mM NaCl interfacial excess charge data and theoretical predictions

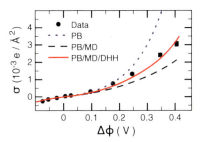

that effectively diverges at $\Delta\phi > 0.3$ V, as shown in Fig. 7.4. This is a clear indication that excluding solvent correlations makes the PB theory ill-conditioned to predict electrostatics at the electrified liquid/liquid interfaces. This conclusion is similar to the one arrived at in Sect. 5.5 with regards to the 10 mM LiCl/5 mM BTPPATPFB system, and later corroborated by x-ray reflectivity measurements. Correcting PB with solvent correlations (PB/MD) through the density functional formalism, with the latter modeled with the help of MD simulations, an idea originally introduced by Luo et al. in [22, 23], gives a well-behaved $\sigma(\phi)$ though it underestimates the data starting at $\Delta\phi > 0.246$ V. In light of the discussion above, this is quite expected, for potentials higher than 0.246 V, the system is becoming strongly correlated with $\Gamma > 2$ and there is a dire need to address electrostatic correlations that allow for more ions at the interface. Remarkable agreement between the PB/MD/DHH model and the data exists at all potentials suggesting that the Debye-Hückel Hole theory properly describes ion correlations and consequently that of ion distributions. Note that *no* adjustable parameters enter any of the theoretical models.

Fitting to the x-ray data involves calculating the electron density profile $\rho(z)$ from the ion density profiles $n(z)$, as described in Sects. 4.2, and 5.5. In addition to probing $d\rho(z)/dz$, XR is also sensitive to interfacial thermal fluctuations (see Sect. 4.1), hence this interfacial roughness constitutes the *only* fitting parameter involved in comparing theoretical density profiles to the data. The fitted roughness values (4.2–4.9 Å) of *all* the simulated XR curves in Figs. 7.5 and 7.6 are within 2 s.t.d (± 0.4 Å) of the interfacial widths predicted from capillary wave theory as shown in Table 7.1, calculated using potential-dependent interfacial tension measurements [28, 29]. This relatively small deviation, between fitted values and a model that has been proven to successfully predict interfacial thermal fluctuations offers strong evidence that the quality of fits shown below between data and PB/MD/DHH (within the available Q_z range) is due to a proper accounting of the system's electron density profile.

In agreement with the macroscopic measurements, the XR data shows that PB/MD correctly predicts ion density profiles at $\Delta\phi < 0.246$ V when the correlation strength is $< 2k_BT$ (see Fig. 7.5), but at higher potentials when stronger TPFB$^-$ correlations are present, it underestimates the interfacial density of the latter, as a result predicting lower reflectivity as illustrated in Figs. 7.6 and 7.7. At larger potentials, we fix the PB/MD roughness to be that of the fitted values obtained for PB/MD/DHH, allowing a meaningful comparison between the predictions of the two models. Note that fitting

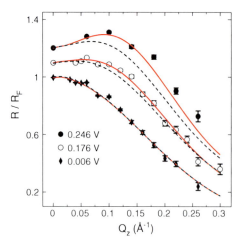

Fig. 7.5 10 mM NaCl XR data and reflectivity predictions of the PB/MD/DHH and PB/MD models for $\Delta\phi \leq 0.246$ V and $\mu^{ion} \leq 2 k_B T$

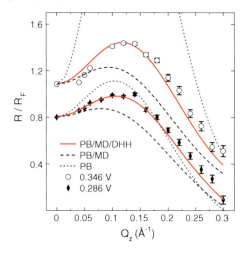

Fig. 7.6 10 mM NaCl XR data and reflectivity predictions of the PB/MD/DHH and PB/MD models for $\Delta\phi \geq 0.286$ V and $\mu^{ion} > 2 k_B T$

Table 7.1 The potential dependent interfacial roughness of the 10 mM NaCl system: capillary wave theory and fits to the X-ray data

	Capillary wave theory (Å)	PB/MD/DHH (± 0.20 Å)	PB/MD (± 0.20 Å)
$\Delta\phi = 0.406$ V	5.23	4.91	4.91
$\Delta\phi = 0.346$ V	4.92	4.53	4.53
$\Delta\phi = 0.286$ V	4.74	4.30	4.30
$\Delta\phi = 0.246$ V	4.61	4.32	4.35
$\Delta\phi = 0.174$ V	4.54	4.21	4.09
$\Delta\phi = 0.006$ V	4.44	4.38	4.40

7.2 Experimental Tests of the PB/MD/DHH Theory

Fig. 7.7 10 mM NaCl XR data and reflectivity predictions of the PB/MD/DHH and PB/MD models at $\Delta\phi = 0.406$ V and $\mu^{ion} \approx 4 k_B T$

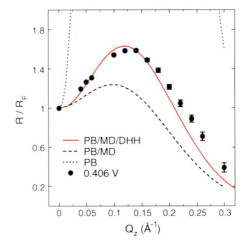

the roughness for PB/MD does not produce substantially better qualitative fits to the data, but results in extremely small roughness values (≈ 3 Å) necessitating large bending rigidities on the order of $100 k_B T$. The PB prediction greatly overestimates the data starting at 0.286 V, indicating that the PB interfacial density is much larger than what occurs in the system. A recurring feature of the PB application to an electrified liquid/liquid interface, as shown in previous systems (Sects. 4.2, 5.5 [30]). Reflectivity curves simulated from the PB/MD/DHH model show good agreement with the reflectivity data at all potentials, capturing the essential trend of the data, and properly describing the ion density profiles. It is noteworthy that PB/MD/DHH produces R/R_F curves higher than those of PB/MD; a direct consequence of the fundamental result that electrostatic correlations allow for more ions to condense on a charged surface.

We note that we ignored TPFB$^-$-Na$^+$ correlations across the interface, due to lack of SC models, to the best of our knowledge, that treat ion–ion correlations across a dielectric discontinuity. Nevertheless, these two ions (and all others in the system) interact in a mean-field way through the electrostatic potential $\phi(z)$. However, such correlations might lead to a more compact double layer [31] and may be responsible for the small deviations at high Q_z between the PB/MD/DHH model and the XR data for the highest potentials. Comparison of the thermodynamic data and XR data of the 10 mM LiCl system against the PB/MD/DHH predictions is shown below in Figs. 7.8 and 7.9, respectively. Also of note is the approximation that f^{sol} is independent of $\Delta\phi$, while current data is well described within this approximation, perhaps the slight deviation between PB/MD/DHH and the data at the highest potential for high Q_z in the case of the 10 mM NaCl data, and the overestimation of the R/R_F peak amplitude for the 10 mM LiCl data at the highest potential, may be an indication of the limitation of this assumption and/or of the DHH model.

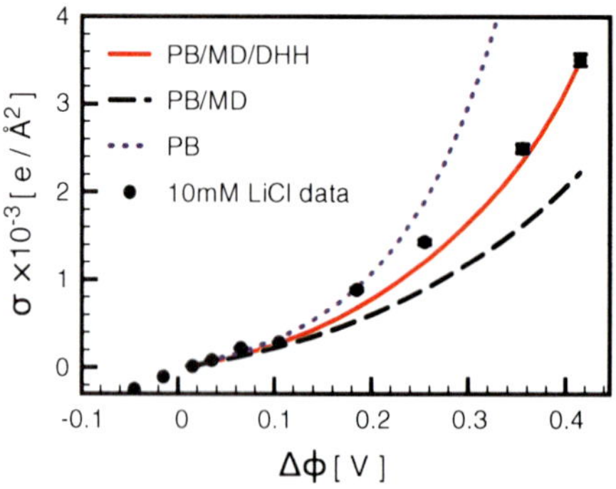

Fig. 7.8 10 mM LiCl interfacial excess charge data and theoretical predictions

Fig. 7.9 10 mM LiCl XR data and reflectivity predictions of the PB/MD/DHH

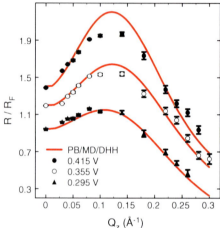

7.4 Conclusion

We presented structural and macroscopic measurements at the electrified liquid/liquid interface that stringently tested the Debye-Hückel-Hole model of ion correlations in a weighted density functional approximation. We found that this model properly accounts for ion correlations up to a correlation strength of $4k_BT$, with density profiles in agreement with XR data and global electrostatic properties in accord with thermodynamic measurements. Therefore, the physical mechanism responsible for the monovalent ion condensation first presented in Chap. 4, and eluded the analysis therein and that of Chap. 5 based on including steric effects of the double layer, is finally identified with strong ion–ion correlations. However, note that the excess chemical potential derived in this section is in qualitative agreement with the phenomenological potential of mean force (Fig. 4.4) found in Chap. 4. Key to the analysis presented in this chapter was the inclusion of specific ion–solvent interactions simulated from MD dynamics that allowed us to quantify the interfacial structure that an ion "sees". The density functional theory presented here is not limited to a liquid/liquid interface and should be applicable in other settings such as ion correlations near charged biomolecules or charged solid surfaces. Moreover, this theory may be extendable to treat dense strongly-coupled systems where the ion–solvent correlations are necessarily a function of density, i.e. $f^{sol}[n_\pm]$.

This comparison between the density profile predictions of a strong coupling limit model and XR measurements is a first in the study of ion correlations near charged surfaces and does offer a significant insight to the study of electrical double layers. A significant contribution of the work presented resides in the conclusion that the rich complexity of interactions found in charged soft matter systems cannot be passed over in empirical investigations of electrostatic ion correlations. Furthermore, The coupling strength Γ for ion correlations in this work is comparable to that for trivalent ions in aqueous solution and we expect that these results will also be relevant for that situation with direct applicability to biophysical settings.

References

1. Rouzina, I., Bloomfield, V.A.: Macroion attraction due to electrostatic correlation between screening counterions. 1. mobile surface-adsorbed ions and diffuse ion cloud. J. Phys. Chem. **100**, 9977 (1996)
2. Shklovskii, B.: Wigner crystal model of counterion induced bundle formation of rodlike polyelectrolytes. Phys. Rev. Lett. **82**(16), 3268–3271 (1999)
3. Grosberg, A.Y., Nguyen, T.T., Shklovskii, B.I.: The physics of charge inversion in chemical and biological systems. Rev. Mod. Phys. **74**, 329 (2002)
4. Moreira, A.G., Netz, R.R.: Binding of similarly charged plates with counterions only. Phys. Rev. Lett. **87**(7), 078301 (2001)
5. Šamaj, L., Trizac, E.: Counterions at highly charged interfaces: From one plate to like-charge attraction. Phys. Rev. Lett. **106**(7), 078301 (2011)
6. Stevens, M., Robbins, M.: Density functional theory of ionic screening: when do like charges attract? Europhys. Lett. **12**, 81–86 (1990)
7. Penfold, R., Nordholm, S., Jonsson, B., Woodward, C.E.: A simple analysis of ion-ion correlation in polyelectrolyte solutions. J. Chem. Phys. **92**(3), 1915 (1990)
8. Nordholm, S.: Simple analysis of the thermodynamic properties of the one-component plasma. Chem. Phys. Lett. **105**, 301 (1984)
9. Groot, R.D.: Ion condensation on solid particles: theory and simulations. J. Chem. Phys. **95**(12), 9191 (1991)
10. Diehla, A., Tamashiro, M.N., Barbosa, M.C., Levin, Y.: Density-functional theory for attraction between like-charged plates. Phys. A **274**, 433–445 (1999)
11. Barbosa, M.C., Deserno, M., Holm, C.: A stable local density functional apporach to ion-ion correlations. Europhys. Lett. **52**, 80 (2000)
12. Kjellander, R., Marcelja, S.: Inhomogeneous coulomb fluids with image interactions between planar surfaces. i. J. Chem. Phys. **82**(4), 2122 (1985)
13. Kjellander, R.: Inhomogeneous coulomb fluids with image interactions between planar surfaces. ii. on the anisotropic hypernetted chain approximation. J. Chem. Phys. **88**(11), 7129 (1988)
14. Kjellander, R., Marcelja, S.: Inhomogeneous coulomb fluids with image interactions between planar surfaces. iii. distribution functions. J. Chem. Phys. **88**(11), 7138 (1988)
15. Kjellander, R., et al.: Double-layer ion correlation forces restrict calcium-clay swelling. J. Phys. Chem. **92**, 6489–6492 (1988)

16. Wernersson, E., Kjellander, R., Lyklema, J.: Charge inversion and ion ion correlation effects at the mercury/aqueous MgSO$_4$ interface: toward the solution of a long-standing issue. J. Phys. Chem. C **114**, 1849–1866 (2010)
17. Wang, W., Park, R.Y., Travesset, A., Vaknin, D.: Ion-specific induced charges at aqueous soft interfaces. Phys. Rev. Lett. **106**(5), 056102 (2011)
18. Martín-Molina, A., Rodríguez-Beas, C., Faraudo, J.: Charge reversal in anionic liposomes: experimental demonstration and molecular origin. Phys. Rev. Lett. **104**(16), 168103 (2010)
19. Debye, P. Hückel, E.: The theory of electrolytes. i. lowering of freezing point and related phenomena. Phys. Z. **24**, 185–206 (1923)
20. Levin, Y.: Electrostatic correlations: from plasma to biology. Rep. Prog. Phys. **65**, 1577 (2002)
21. Baus, M., Hansen, J.-P.: Statistical mechanics of simple coulomb systems. Phys. Rep. **59**(1), 1–94 (1980)
22. Luo, G., Malkova, S., Yoon, J., Schultz, D.G., Lin, B., Meron, M., Benjamin, I., Vanysek, P., Schlossman, M.L.: Ion distributions at the nitrobenzene-water interface electrified by a common ion. J. Electroanal. Chem. **593**, 142–158 (2006)
23. Luo, G., Malkova, S., Yoon, J., Schultz, D.G., Lin, B., Meron, M., Benjamin, I., Vanysek, P., Schlossman, M.L.: Ion distributions near a liquid-liquid interface. Science **311**, 216–218 (2006)
24. Wick, C.D., Dang, L.X.: Molecular dynamics study of ion transfer and distribution at the interface of water and 1,2-dichloroethane. J. Phys. Chem. C **112**(3), 647–649 (2008)
25. Hou, B.: Ion distributions at electrified liquid-liquid interfaces: an agreement between X-ray reflectivity analysis and macroscopic measurements. Doctoral dissertation, University of Illinois at Chicago (2011)
26. Tarazona, P.: Free-energy density functional for hard spheres. Phys. Rev. A **31**(4), 2672 (1985)
27. Curtin, W., Ashcroft, N.: Phys. Rev. A **32**, 2909 (1985)
28. Mitrinovic, D.M., Tikhonov, A.M., Li, M., Huang, Z., Schlossman, M.L.: Noncapillary-wave structure at the water-alkane interface. Phys. Rev. Lett. **85**, 582 (2000)
29. Buff, F.P., Lovett, R.A., Stillinger, F.H.: Interfacial density profile for fluids in the critical region. Phys. Rev. Lett. **15**, 621 (1965)
30. Laanait, N., Yoon, J., Hou, B., Vanysek, P., Meron, M., Lin, B., Luo, G., Benjamin, I., Schlossman, M.: Communications: monovalent ion condensations at the electrified liquid/liquid interface. J. Chem. Phys. **132**, 171101 (2010)
31. Torrie, G., Valleau, J.: Double layer structure at the interface between two immiscible electrolyte solutions. J. Electroanal. Chem. **206**, 69–79 (1986)

Appendices

Chapter 2

Atomic coordinates of the compound BTPPATPFB, as obtained from x-ray powder diffraction measurements, in XYZ format:

TPFB-45

```
B    -0.64230    16.75360    10.21640
C    -1.19890    17.19450    11.70850
C    -1.41100    18.54150    11.98680
C    -1.85180    19.03560    13.18410
C    -2.10970    18.18190    14.21790
C    -1.89230    16.84400    14.02350
C    -1.42670    16.38920    12.80740
C    -0.61550    15.11990     9.96130
C    -1.77080    14.37000    10.14820
C    -1.87350    13.02230     9.89750
C    -0.79810    12.35080     9.39470
C     0.35370    13.03190     9.14530
C     0.42780    14.38770     9.42960
C    -1.62280    17.27260     8.97620
C    -2.91470    17.75210     9.08050
C    -3.69490    18.09580     7.98580
C    -3.21040    17.93750     6.72810
C    -1.95140    17.43130     6.56100
C    -1.20680    17.11060     7.67090
C     0.86690    17.44220    10.20190
C     1.27770    18.56520     9.50960
C     2.52220    19.14170     9.63200
C     3.44530    18.60520    10.48530
C     3.09400    17.50720    11.21720
C     1.83900    16.96750    11.07190
F    -1.17660    19.45660    11.02000
F    -2.02700    20.35840    13.35260
F    -2.54430    18.63970    15.39700
F    -2.13100    15.97960    15.02490
F    -1.17920    15.05910    12.75330
F    -2.88700    14.97800    10.60150
F    -3.02930    12.36720    10.12640
F    -0.86980    11.03310     9.13390
F     1.42460    12.40170     8.63680
F     1.61480    14.96290     9.13930
F    -3.50660    17.91890    10.27420
F    -4.93430    18.58710     8.17620
F    -3.95600    18.27110     5.65720
F    -1.46270    17.24690     5.32330
F     0.01750    16.60420     7.43060
F     0.45240    19.19330     8.64090
F     2.83910    20.23570     8.91350
F     4.66400    19.15350    10.61020
```

Appendices

#BTPPA+
69

P	2.20030	2.23050	4.72620
P	0.37260	-0.02010	3.96690
N	1.44510	1.12680	3.89880
C	2.38690	3.68210	3.69340
C	2.35820	3.53630	2.31620
H	2.22680	2.69600	1.94180
C	2.52130	4.62450	1.50610
H	2.50600	4.52270	0.58260
C	2.70620	5.86110	2.05200
H	2.80110	6.60140	1.49660
C	2.75440	6.02390	3.41840
H	2.89610	6.86840	3.78110
C	2.59350	4.93410	4.24820
H	2.62320	5.03920	5.17190
C	3.83450	1.66440	5.22370
C	4.90120	2.53090	5.31700
H	4.80340	3.41910	5.05820
C	6.11200	2.07990	5.79390
H	6.82930	2.66830	5.85550
C	6.26790	0.78810	6.17600
H	7.08970	0.49640	6.50120
C	5.22680	-0.07990	6.08140
H	5.33510	-0.96750	6.34020
C	4.00950	0.35320	5.60450
H	3.30040	-0.24440	5.53980
C	1.36320	2.75970	6.22820
C	1.81460	2.39340	7.48320
H	2.60460	1.90990	7.57160
C	1.09170	2.74610	8.60200
H	1.39160	2.49190	9.44400
C	-0.06470	3.47020	8.48200
H	-0.54600	3.70630	9.24190
C	-0.51410	3.84840	7.24240
H	-1.29800	4.34130	7.16430
C	0.19150	3.49950	6.11520
H	-0.11550	3.75920	5.27610
C	0.62000	-1.10180	2.55610
C	-0.42420	-1.74630	1.95700
H	-1.29540	-1.57470	2.23070
C	-0.18650	-2.64810	0.95040
H	-0.89990	-3.10160	0.56200
C	1.07680	-2.88570	0.51470
H	1.22470	-3.48210	-0.18160

Chapter 4

We derive the functional derivative of the electrostatic self-energy, $|\nabla\phi(\mathbf{r})|^2$ used in the derivation of the sterically modified Poisson-Boltzmann Eq. (4.67).

$$\frac{\delta|\nabla\phi(\mathbf{r}')|^2}{\delta\phi(\mathbf{r})} = 2\frac{\delta}{\delta\phi(\mathbf{r})}(\nabla\phi(\mathbf{r}')).\nabla\phi(\mathbf{r}') \qquad (1)$$

$$= 2\nabla\frac{\delta\phi(\mathbf{r}')}{\delta\phi(\mathbf{r})}.\nabla\phi(\mathbf{r}') \qquad (2)$$

$$= 2\nabla\delta(\mathbf{r}' - \mathbf{r}).\nabla\phi(\mathbf{r}') \qquad (3)$$

$$= -2\delta(\mathbf{r}' - \mathbf{r})\nabla^2\phi(\mathbf{r}') + 2\nabla.(\delta(\mathbf{r}' - \mathbf{r})\nabla\phi(\mathbf{r}')) \qquad (4)$$

$$= -2\delta(\mathbf{r}' - \mathbf{r})\nabla^2\phi(\mathbf{r}') \qquad (5)$$

We apply the chain rule in the 1st line. In the 2nd line, we exchange the order of the functional derivative and the gradient with respect to $\mathbf{r}'$. In the 4th line, we use the chain rule once more. Since the whole expression appears under an integral, the second term on the right of line 4th vanishes once we use the divergence theorem, giving the desired expression.

In the following, we take an in depth look into PBSolve. We include the INPUT, DEFINITIONS, MODELS sections used for a 1-phase system. For a 2-phase system, the main difference is in the Code (algorithm) section used to solve the double layer model, which we also reproduce.

PBSolve for electrolyte in contact with charged surface

Version: 4
Author: N. Laanait
First Modified: 10/02/2009
Last Modified: 10/22/2010
Note: the code is fully parallel. Uses a max of 8 processors. But works fine for < 8.

```
ClearAll["Global`*"]
AbortKernels[];
```

- **INPUT**
  - Setting the working directory and name of output files
    ```
    (*workdirectory=SetDirectory["~/SPB_Project/test"];*)
    workdirectory = SetDirectory[
        "/Users/nlaanait/work/Programming/double_layer_codes/SPB_Project/GSPB/divalent"];
    dataphi = "Phi_1da.dat";
    phiplot = "Phi_1da.dat";
    datarho = "Rho_1da.dat";
    plotrho = "Rho_1da.pdf";
    ionlabel[1] = "Na+";
    ionlabel[2] = "Cl-";
    ionlabel[3] = "Ba2+";
    ionlabel[4] = "OH-";
    ```

Appendices

- **Input of physical properties**

```
(* Temperature in K *)
T = 300;
(* Dielectric constant *)
ϵ = 80;
(* number of ions*)
nions = 3;
(* ion i Bulk concentration in Molarity *)
cb[1] = 0.001;
cb[2] = cb[1] + z[3] cb[3];
cb[3] = 0.001;
cb[4] = 3 * 10^-6;
(* ion i valency and charge *)
z[1] = 1;
z[2] = -1;
z[3] = 2;
z[4] = -1;
(* ion i diameter for SBP equation*)
d[1] = 1;
d[2] = 3.6;
d[3] = 8;
d[4] = 2.8;
(* equation to solve: PB, SPB, LDA, WDA *)
eq = "SPB";
(*type of boundary condition,
BVP= boundary value problem, specify ΔV in Volts.
    IVP= initial value problem, specify σ in electrons/Å². *)
bctype = "BVP";
ΔV = -0.35;
σ = 10^-3;
(* Excess chemical potential to use in LDA model *)
chemex = "OCP"; (* List of ions with μ_ex *)
chemexi = {1}; (* Trial solution for LDA, file *)
trialfile = "Rho_lda.dat";
column = 2; (* Column in the file of the trial solution concentration *)
code = "seq" (* method to solve Poisson's equation: seq (sequential),
  prl (parallel) *);
```

- **CONSTANTS & DEFINITIONS**

- **MODELS OF ION DISTRIBUTIONS**

- **PB:** Boltzmann Distribution in the Electrostatic Potential

```
cPB[i_] = cb[i] Exp[-z[i] χ[x]];
```

- **GSPB:** Generalized Sterically modified Poisson-Boltzmann equation

- **POISSON EQUATION**

```
ClearAll[χ, phi]

(* Choosing which distribution to use *)
c[i_] := Which[eq == "PB", cPB[i],
               eq == "SPB", cSPB[i]
              ];

(*Poisson's equation for planar geometry,
x-dir normal to the charged surface. [x]=Å, [ϕ]=Volts. *)
Poisson := χ''[x] == -α Sum[z[i] c[i], {i, 1, nions, 1}];
```

- **BOUNDARY CONDITIONS**

  ```
  Chi = ΔV * θ;
  Chis = -σ * 16.0217646 * θ / (10^10 ϵvac ϵ);
  charge = σ / Chis χ'[0];
  BCS = {Which[bctype == "BVP", χ[0] == Chi, bctype == "IVP", χ'[0] == Chis], χ'[10 dbl] == 0};
  ```

- **Code**
- **PLOTS & DATA FILES**
- **Code**

  PBI (II) is the Poisson-Boltzmann equation in phase I (II).
  BCSI (II) are the boundary conditions in phase I (II).
  Below we define the equation to be solved by FindRoot[ ].
  χI0 is ϵI E$_I$ (0) in code units, where E is the electric field

  ```
  χI0[chi0_?NumberQ] :=
    ϵI First[χ'[0] /. NDSolve[{PBI, BCSI}, χ, {x, 0, 10 dblI}, MaxSteps → ∞,
        InterpolationOrder → All, Method →
          {"Shooting", "StartingInitialConditions" → {χ'[5 dblI] == 0, χ[5 dblI] == 0}}
        ] // Quiet];

  χII0[chi0_?NumberQ] :=
    ϵII First[χ'[0] /. NDSolve[{PBII, BCSII}, χ, {x, -10 dblII, 0}, MaxSteps → ∞,
        InterpolationOrder → All, Method →
          {"Shooting", "StartingInitialConditions" → {χ'[-5 dblII] == 0, χ[-5 dblII] == 0}}
        ] // Quiet];
  ```

  (* We find α (chi0) such that the Electric field BC is satisfied *)

  ```
  α = Flatten[FindRoot[χII0[chi0] == χI0[chi0], {chi0, ΔV θ / 2}]];
  ```

  (* Given α we use NDSolve[ ] to solve PBI and PBII *)

  ```
  SolI = NDSolve[{PBI, BCSI} /. α, χ, {x, 0, 10 dblI},
      MaxSteps → ∞, InterpolationOrder → All, Method → {"Shooting",
        "StartingInitialConditions" → {χ'[4 dblI] == 0, χ[4 dblI] == 0}}] // Quiet;

  SolII = NDSolve[{PBII, BCSII} /. α, χ, {x, -10 dblII, 0},
      MaxSteps → ∞, InterpolationOrder → All, Method → {"Shooting",
        "StartingInitialConditions" → {χ'[-4 dblII] == 0, χ[-4 dblII] == 0}}] // Quiet;
  ```

Chapter 5

Partial Charges of TPFB$^-$ computed using CHELPG algorithm without any constraints.

CHELPG computation with TPFB- crystal structure geometry.

Partial Charges Model: MP2. Basis: (aug-cc-pVDZ)
ESP

```
 1 B   0.001  0.001  0.001   2.623403
 2 C  -0.112  0.106  0.060  -1.084941
 3 C  -0.174  0.195 -0.027   0.296826
 4 C  -0.267  0.288  0.009   0.147038
 5 C  -0.305  0.299  0.140   0.224324
 6 C  -0.246  0.217  0.232   0.061726
 7 C  -0.150  0.126  0.191   0.436759
 8 C   0.053 -0.111  0.112  -1.219071
 9 C  -0.039 -0.191  0.178   0.502724
10 C  -0.006 -0.291  0.267  -0.017463
11 C   0.125 -0.319  0.290   0.314883
12 C   0.221 -0.248  0.225  -0.010619
13 C   0.184 -0.146  0.138   0.525233
14 C  -0.058 -0.097 -0.120  -1.044380
15 C  -0.190 -0.117 -0.154   0.419032
16 C  -0.231 -0.206 -0.252   0.072353
17 C  -0.140 -0.281 -0.318   0.217423
18 C  -0.008 -0.269 -0.286   0.125707
19 C   0.029 -0.179 -0.189   0.321394
20 C   0.121  0.104 -0.049  -0.952022
21 C   0.154  0.139 -0.178   0.327699
22 C   0.248  0.233 -0.212   0.084937
23 C   0.316  0.301 -0.113   0.268478
24 C   0.286  0.273  0.017   0.056388
25 C   0.191  0.178  0.046   0.351624
26 F  -0.142  0.191 -0.159  -0.213461
27 F  -0.322  0.369 -0.083  -0.201376
28 F  -0.397  0.389  0.178  -0.193644
29 F  -0.280  0.226  0.361  -0.188965
30 F  -0.093  0.055  0.291  -0.207123
31 F  -0.171 -0.171  0.157  -0.248694
32 F  -0.102 -0.362  0.328  -0.178731
33 F   0.160 -0.417  0.375  -0.200890
34 F   0.351 -0.274  0.244  -0.185833
35 F   0.287 -0.082  0.078  -0.221062
36 F  -0.289 -0.049 -0.093  -0.202300
37 F  -0.362 -0.216 -0.282  -0.193068
38 F  -0.179 -0.368 -0.414  -0.192552
39 F   0.084 -0.345 -0.349  -0.194811
40 F   0.161 -0.173 -0.161  -0.221671
41 F   0.094  0.080 -0.284  -0.195239
42 F   0.275  0.261 -0.341  -0.190966
43 F   0.408  0.393 -0.145  -0.203470
44 F   0.349  0.338  0.116  -0.182601
45 F   0.167  0.156  0.177  -0.232996
                              ----------------
                              -1.000000
```

Dipole moment   0.105382 0.093133 0.093133

Quadrupole moment Qxx -2.573069 -2.497597 -2.497597
                   Qyy   4.316082 4.159055 4.159055
                   Qzz -1.743014 -1.661458 -1.661458

Chapter 6

Mathematica Code to solve the ion density profiles using the WDA of a DHH plasma. The Poisson equation solver used is `PBSolve`, see Chap. 4 appendix for its outline.

- **INPUT**
- **CONSTANTS & DEFINITIONS**
- **MODELS OF ION DISTRIBUTIONS**

- **PB:** Boltzmann Distribution in the Electrostatic Potential
- **Solvent PMF**
- **WDA:** Weighted Density Approximation of DHH plasma

```
(*Importing test file*)

If[model == "WDA",
  data = Import[trialfile];
  ctrial[x_] =
    Interpolation[Table[{data[[i, 1]], data[[i, 3]]},
       {i, 3, Length[data]}], Method → "Spline"][x];
  (*ctrial[x_]=Interpolation[Table[{-x,cbII[1]},{x,0,10 dblII}]][
     x];*)

  (* DHH Plasma *)
  κ := Sqrt[4 * Pi * blII * (NA * 10^-27) * ρ];
  ω := (1 + 3 blII * κ)^(1/3);
  fOCP :=
    1/(4 β) (1 - ω^2 + Log[(ω^2 + ω + 1)/3]) + (2 Pi)/(3 Sqrt[3]) -
      2/Sqrt[3] ArcTan[(2 ω + 1)/Sqrt[3]]);
  δfOCP = D[fOCP, ρ];
  scorr[x_] =
    Interpolation[Table[{x, (ω/κ - 1/κ) /. ρ → ctrial[-x]},
       {x, 0, 10 dblII}], Method → "Spline"][-x];
  cw[x_] = Interpolation[
      ParallelTable[{x,
          Which[x > 5 dblII, ctrial[-x], x ≤ 5 dblII,
            3/(2 (scorr[-x])^3)
              NIntegrate[ctrial[-y] (scorr[-x] - Abs[x - y])^2,
```

Appendices

```mathematica
            {y, Max[0, x - scorr[-x]], x + scorr[-x]},
            MaxRecursion → 20]]},
          {x, 0, 9 dblII, 1}],
        Method → "Spline"][-x];
    intfOCP[x_] =
     Interpolation[ParallelTable[
        {x, NIntegrate[cw[-y] * (NA * 10^-27) (δfOCP /. ρ → cw[-x]),
           {y, 0, 10 dblII}]}, {x, 0, 10 dblII, 10}], Method → "Spline"][
       -x];
    μexDHH[x_] =
     Interpolation[Table[{x, intfOCP[-x] + fOCP /. ρ → cw[-x]},
        {x, 0, 10 dblII}], Method → "Spline"][-x]
   ];

cWDAII[i_] =
  cbII[i]
    Exp[-zII[i] χ[x] - β (μexDHH[x] - μexDHH[-10 dblII])
      KroneckerDelta[1, WionflagII[i]] -
      WsolII[i][x] KroneckerDelta[1, WsolflagII[i]]];
cWDAI[i_] = cPMFI[i];
```

- **POISSON's Equation**
- **Boundary Conditions**
- **Code**

- PBSolve
- PBSolve self-consistent

```mathematica
  iter := Do[
    Print["Iteration # =", q];
    If[
     q == 0,
     (* First iteration step*)
     rho[0][x_] = ctrial[x];
     scorr[x_] = Interpolation[
           Table[{x, (ω/κ - 1/κ) /. ρ → rho[0][-x]},
        {x, 0, 10 dblII}]
           , Method → "Spline"][-x];
     cw[x_] = Interpolation[
```

```
              ParallelTable[{x,
                      Which[x > 5 dblII, rho[0][-x],
                                    3
  x ≤ 5 dblII,  ─────────────
                            2 (scorr[-x])^3
              NIntegrate[rho[0][-y] (scorr[-x] - Abs[x - y])^2,
              {y, Max[0, x - scorr[-x]], x + scorr[-x]},
           MaxRecursion → 20]]},
                  {x, 0, 9 dblII, 1}],
       Method → "Spline"][-x];
intfOCP[x_] =
 Interpolation[ParallelTable[
    {x, NIntegrate[cw[-y] * (NA * 10^-27) (δfOCP /. ρ → cw[-x]),
      {y, 0, 10 dblII}]}, {x, 0, 10 dblII, 10}],
   Method → "Spline"][-x];
μexDHH[x_] =
 Interpolation[Table[{x, intfOCP[-x] + fOCP /. ρ → cw[-x]},
    {x, 0, 10 dblII}], Method → "Spline"][-x];


cWDAII[i_] =
 cbII[i]
  Exp[-zII[i] χ[x] - β (μexDHH[x] - μexDHH[-10 dblII])
     KroneckerDelta[1, WionflagII[i]] -
     WsolII[i][x] KroneckerDelta[1, WsolflagII[i]]];
cWDAI[i_] = cPMFI[i],


(* Subsequent iterations with mixing *)
     rho[q][x_] = (1 - γ) rhosol[q - 1][x] + γ rho[q - 1][x];

scorr[x_] =
 Interpolation[Table[{x, (ω/κ - 1/κ) /. ρ → rho[q][-x]},
    {x, 0, 10 dblII}], Method → "Spline"][-x];
   cw[x_] = Interpolation[
              ParallelTable[{x,
                      Which[x > 5 dblII, rho[q][-x],
                                    3
  x ≤ 5 dblII,  ─────────────
                            2 (scorr[-x])^3
              NIntegrate[rho[q][-y] (scorr[-x] - Abs[x - y])^2,
              {y, Max[0, x - scorr[-x]], x + scorr[-x]},
```

```
          {y, Max[0, x - scorr[-x]], x + scorr[-x]},
         MaxRecursion → 20]]},
            {x, 0, 9 dblII, 1}],
        Method → "Spline"][-x];

intfOCP[x_] =
 Interpolation[ParallelTable[
    {x, NIntegrate[cw[-y] * (NA * 10^-27) (δfOCP /. ρ → cw[-x]),
       {y, 0, 10 dblII}]}, {x, 0, 10 dblII, 10}],
    Method → "Spline"][-x];

μexDHH[x_] =
 Interpolation[Table[{x, intfOCP[-x] + fOCP /. ρ → cw[-x]},
    {x, 0, 10 dblII}], Method → "Spline"][-x];

cWDAII[i_] =
  cbII[i]
   Exp[-zII[i] χ[x] - β (μexDHH[x] - μexDHH[-10 dblII])
      KroneckerDelta[1, WionflagII[i]] -
      WsolII[i][x] KroneckerDelta[1, WsolflagII[i]]];
cWDAI[i_] = cPMFI[i]

  ];

(* Solving Poisson's equation with rho[q]*)

SystemOpen[Export["scorr.pdf",
   Plot[scorr[x], {x, -2 dblII, 0}, PlotStyle → Thick,
    PlotRange → All]]];
PoissonI := χ''[x] == -αI Sum[zI[i] cWDAI[i], {i, 1, nionsI}];
PoissonII := χ''[x] == -αII Sum[zII[i] cWDAII[i], {i, 1, nionsII}];
If[q == 0, guess = ΔVθ/2, guess = α[q - 1][[1, 2]]];
pbsolve;
rhosol[q][x_] = cWDAII[2] /. SolII [[1]];

(* Checking Convergence *)
If[
    q ≠ 0,

  Tol = Max[Table[{Abs[rhosol[q][-x] - rhosol[q - 1][-x]]},
    {x, 0, 3 dblII}]];
    Print["Tolerance = ", Tol];
    SystemOpen[
      Export["ccomp.pdf",
```

```
        Export["ccomp.pdf",
   LogPlot[{rhosol[q][x], rhosol[q-1][x]}, {x, -4 dblII, 0},
    PlotRange → All, PlotStyle → Thick,
    AxesLabel → {"z [Å]", "c(z) [M]"},
    PlotLabel → "Convergence Not Reached, ρ_before-ρ_now =  " <>
       ToString[FortranForm[Raund[Tol, 7]]] <> "M",
    PlotLegend → {"ρ-now", "ρ-before"},
    LegendPosition → {1.1, -0.4}]]];

   SystemOpen[
      Export["mucomp.pdf",
   Plot[{β μexDHH[x] - β μexDHH[-10 dblII],
     WsolII[2][x] + β μexDHH[x] - β μexDHH[-10 dblII]},
    {x, -4 dblII, 0}, PlotRange → All, PlotStyle → Thick,
    PlotLegend → {"Wion", "Wsol+Wion"}, LegendPosition → {1.1, 0},
    AxesLabel → {"z [Å]", "W(z)[kT]"}]]];

   If[Tol ≤ 10^-6,
            Print["Convergence Reached, Error in ρ(z)(M) = ",
   Tol];
            Break[]
      ]
 ]

 , {q, startiter, maxiter}

]
```

- **Solution**

- **PLOTS & DATA FILES**